STEVE POPE is a teacher and keen
runner who hopes to retire soon to
spend more time with his debts.

SO THAT'S WHY THEY CALL IT GREAT BRITAIN

How One Tiny Country Gave So Much To The World

Steve Pope

Monday Books

© Steve Pope 2009

First published in the UK in 2009 by Monday Books

The right of Steve Pope to be identified as the Author of
this work has been asserted by him in accordance with the
Copyright, Designs and Patents Act 1988

A CIP catalogue record for this title is available from the
British Library

ISBN: 978-1-906308-06-3

Typeset by Andrew Searle

Printed and bound by CPI Cox and Wyman,
Reading RG1 8EX

www.mondaybooks.com
info@mondaybooks.com

'A report by the Ministry of International Trade and Industry, the Japanese governmental organisation, found that over 40 per cent of major discoveries during the past 50 years came from the UK.'

The Guardian

Foreword

TO mapmakers, 'Great Britain' is simply the largest of the many islands which make up the United Kingdom. To the French, the name merely distinguishes 'Grande Bretagne' from Brittany. For many Britons, though, it has more romantic allusions – to the history of our country, and the achievements of its people.

Britain hasn't always been great, and grand civilisations and empires existed long before it could be considered remotely so. Other modern nations are certainly great, too: there are wealthier and more powerful countries than Britain, and great scientists and philosophers and artists and thinkers can be found in all corners of the globe.

But for all that, Britain is undeniably 'Great' – perhaps the Greatest nation the world has ever seen. It covers less than one per cent of the world's land mass, but its achievements are huge, enormous, stunning. This book attempts to describe some of them.

Primarily, it is not about institutions but people – the doers and inventors of *things*. 'Inventor' is a loaded term, of course: while I'm talking about the British, people from other countries often contributed to our creations. To paraphrase Isaac Newton, all great people stand on the shoulders of others. Similarly, it's a westerner's view: although I describe Captain Cook as being the first man to discover Hawaii, I know that Polynesians beat him to it by several thousand years. But what makes Cook 'great' is that he travelled all the way to them before *they* thought of travelling to *us* (otherwise, our history would be very different).

Some characters – like Cook and Newton, and William Shakespeare and Charles Darwin – are singled out. Many others – like Lord Kelvin, or James Clerk Maxwell – receive only a line or two, when their contributions to science and the world at large are worth 1,000 pages at least. But then, this is not intended to be a definitive work (though if you would like to have anything added to future editions of this book, please email the publishers at info@mondaybooks.com. The idea is to cover as much as possible in as light and informative a way as possible – to give younger readers some idea of the vast breadth of our great country's achievements, and to remind everyone else of what they learned at school but have long forgotten.

I take it as read that readers understand (even if only at a basic level) concepts like radioactivity, electricity and gravity. All financial comparisons are from http://measuringworth.com and compare average earnings unless stated. References to the 'Royal Society' are to the Royal Society of London for the Improvement of Natural Knowledge, founded in 1660 and the world's oldest scientific society. I've used many sources; any errors are mine.

Steve Pope, 2009.

THE AEROPLANE

WE all know the story of the Wright brothers – the Americans who flew the world's first powered aeroplane in 1903. It's a fascinating and romantic tale which has spawned books and films and turned the pair into US icons.

There's just one problem: the real story of flight starts on *this* side of the Atlantic.

Sir George Cayley (1773-1857) is the real 'Father of Aerodynamics'. In 1849, he launched a glider at Brompton, Yorks – it was **the world's first flight**. Wisely, he employed an unnamed 10-year-old boy as his test pilot. (Sir George, a Scarborough baronet and MP, was a prodigious inventor who also worked on self-righting lifeboats, caterpillar tracks and seat belts).

John Stringfellow was a tool maker from Sheffield who moved to Chard in Somerset to work in the lace industry. Slightly eccentrically – it was 1842, and airliners were many decades away – he dreamed of setting up the world's first international airline. With engineer William Henson, he set up the Aerial Transit Company and commissioned brochures which showed aeroplanes carrying passengers on sight-seeing trips over exotic locations like the Pyramids. You can't fault them for thinking big, but unsurprisingly they failed to attract any serious financial backers. The disillusioned Henson emigrated to America, but Stringfellow (1799-1883) continued undeterred, working out optimum wing shapes and materials, and calculating the surface area of wing needed per pound of weight to produce lift. He built a working aircraft, which had a three-metre wing made from silk with cane struts, featuring a rigid leading edge combined with a looser trailing edge, and a steam engine which powered two propellers. However, it was so delicate that outdoor flights proved impossible – the silk became heavy with atmospheric moisture and gusts of wind could be disastrous – so early attempts at flight took place inside a large silk mill in Chard. Finally, in 1848, Stringfellow's machine took to the air, unmanned. Although it travelled less than 10 metres, this was **the world's first *powered* flight**.

Perhaps disheartened after his years of labour for little reward, Stringfellow – the modern-day lapdance guru Peter is descended from him – put his machine on ice. A model of his aeroplane is on display at the Science Museum.

The quest for manned flight remained. In 1899, the British engineer and glider pilot **Percy Pilcher** (1867-1899) came very close to being a household name and scooping the Wright brothers by several years. Pilcher had earlier designed and built a glider called *Hawk*, and had set a gliding distance record of 250 metres near Eynsford in Kent. Then he set about developing a powered aeroplane, and settled on the idea of a three-winged triplane (multiple wings give extra lift without the huge increase in weight that a single wing of the same total area would need.) Power came from a small internal combustion engine. He had arranged to demonstrate his triplane to the public on September 30, 1899, at Stanford Hall near Rugby. Unfortunately, it was not ready and, so as not to disappoint the crowd, he decided to fly *Hawk* instead. Conditions were very blustery, and the glider's tail snapped off in flight, sending it crashing to the ground, killing Pilcher. In 2002, BBC's *Horizon* commissioned the construction of a replica of his triplane at Cranfield University. It flew for over a minute and was a better machine than the Wright Brothers' 'Flyer'. A monument to Pilcher was built on the spot where he crashed; the original *Hawk* is at the Museum of Flight in East Lothian, and a replica can be seen at Stanford Hall.

AMERICA, DISCOVERY OF

THE Spanish explorer Christopher Columbus is said to have discovered America in 1492. Actually, he landed in the Caribbean and it was the Spanish *conquistador* Juan Ponce de León who made the first documented European arrival on the mainland of the future USA on April 2, 1513.

However, a Welshman may have beaten him to it by almost 350 years. Some historians believe that **Prince Madoc ab Owain**

Gwynedd, born the illegitimate son of a Welsh king in 1150, set sail in search of new lands in 1169 and landed near what is now Mobile, Alabama, in 1170. Delighted by what he had found, he is said to have returned to his homeland and then gone on a second voyage with ten boats-full of friends and family who were keen to see this wondrous new country.

Richard Deacon, author of *Madoc and the Discovery of America*, writes that 'he went up the Alabama River and then disappeared in the wilds of Tennessee'. In 1738, the Canadian explorer Pierre Gaultier de Varenne found 'fair-skinned Indians… who spoke some Welsh words and put sentences together in the way Welsh people do'. Called the 'Mandans', de Varenne's detailed notes show that they were white and lived in towns and villages laid out in streets and squares. They also had pre-colonial forts – one was said to be almost identical to Dolwyddelan Castle in Gwynedd, Madoc's birthplace in Wales.

Furthermore, a document written by Governor John Seiver of Tennessee in 1810 recounts a conversation he had with a 90-year-old Cherokee Indian chief called Oconosoto. In the document – which is stored at the Georgia Historical Commission – Seiver writes that he asked the chief who had built these forts. Oconosoto replied that his grandfather had said that they were a 'people called Welsh, and that they had crossed the Great Water and landed first near the mouth of the Alabama River near Mobile'.

It would be nice, of course, to speak to surviving descendants of the Mandan Indians themselves. Sadly, they were wiped out by a smallpox epidemic in 1837. A memorial tablet at Port Morgan, Mobile Bay, laid in 1953, reads: 'In memory of Prince Madog (*sic*), a Welsh explorer, who landed on the shores of Mobile Bay in 1170 and left behind, with the Indians, the Welsh language.'

(Confusingly, some experts claim that *another* Welsh prince, also called Madoc, got there first in 526AD – there are also claims that the Icelander Leif Ericson found America around the year 1000AD.)

AMERICA, CREATION OF

WHOEVER discovered it, the USA – undoubtedly the most powerful nation the world has ever seen – would not exist as it is, were it not for us. We can argue about whether that's a good thing or not, of course!

In the late 1500s, with 'America' now known, **Sir Walter Raleigh** (or Ralegh, 1554-1618) organised a series of voyages. He wanted to seize the riches of the New World and to use it as a base from which to attack the Spanish, who were trading and settling in and around South America, and with whom we were often at loggerheads.

His expeditions were not exactly a roaring success. The first saw his ship run out of supplies after only two days at sea. The second, in 1585, led to a colony being established on the Island of Roanoke off the coast of what is now North Carolina – then named 'Virginia' in honour of the 'virgin queen', Elizabeth I. There were some skirmishes with the natives and the island was abandoned after a year, with a skeleton party of 15 men left behind.

Raleigh's third expedition, in 1587, was led by one John White. White took 121 people, mainly families from Devon, with instructions from Sir Walter to land further to the north and establish a settlement which he modestly instructed them to name 'Raleigh'. They went back to Roanoke first, but on arrival they found no sign of the 15 men from the previous colony. That was ominous but, undeterred, they decided to stay put. White returned to England for supplies, leaving behind the colonists – who included his daughter Eleanor and his new-born granddaughter Virginia Dare (the first English child born in the Americas). He intended to return in three months, but unfortunately war against Spain was about to break out – the Armada would attempt to invade Britain the following year – and all ships were needed for the military effort. Thus, he was not able to return to Roanoke until 1590 and when his ships finally landed – on August 18, his granddaughter's third birthday – they found the settlement deserted.

White had agreed with the settlers that if they had to leave the island they would carve their destination into a tree. He found the word 'Croatoan' etched into a fence post, which suggested they had moved to nearby Croatoan Island. He wanted to sail there to search, but his men – near mutinous in the face of an approaching storm – refused to go. They returned to England, and it's not hard to imagine the anguish the poor man must have felt as he sailed away, knowing he was abandoning forever his daughter and granddaughter to their fates.

(Raleigh is also famous for having lain down his cloak in a puddle so that Queen Elizabeth's feet stayed dry, and for introducing the potato and tobacco to Britain, though the cloak and potato stories are probably untrue. Despite his undoubted services to his country, he was thrown in the Tower of London by King James I in 1603 for supposedly plotting to overthrow him. He used his time to write a *Historie of the World*, but only got as far as 130 BC before being beheaded in 1618.)

The first successful colony was established at what is now Jamestown, Virginia, in 1607. Such was the hardship faced by the settlers that, of the initial 104 colonists, only 38 were still alive when the first supply ship got there from Britain just eight months after they had arrived. By 1609, the population had grown to 600 – but 540 of *those* died from starvation or were killed by 'Red Indians' over that winter. By June 1610, they had decided to abandon the colony and sail back to Britain, but a supply ship arrived in the nick of time and they stayed.

Perhaps the most famous settlement was that at Plymouth, Massachusetts, set up by the **Pilgrim Fathers** who sailed aboard the *Mayflower* in 1620. Once again, dozens of them succumbed to scurvy, tuberculosis and pneumonia *en route* and they had to fight tooth and nail to establish themselves on arrival. They had no GPS to navigate by, no satellite telephone to call for help if things went wrong: they were truly alone.

The following century saw the British drive the French, Spanish and Dutch out of the New World ('New Netherland' was renamed New York State and 'New Amsterdam' New York City in the 1660s and 1670s) and by 1732 there were 13 major British colonies in America. These would eventually form the basis of the following states: New Hampshire, Massachusetts, Maine, Rhode Island, Connecticut, New York, Vermont, New Jersey, Pennsylvania, Delaware, Virginia, Maryland, Kentucky, West Virginia, North Carolina, South Carolina, Tennessee and Georgia.

Britain's rule did not last very long. Although 'Americans' were taxed by us they had no representation in our Parliament and this caused growing anger and resentment. There were a number of flashpoints, among them the shooting of 11 demonstrators by British 'Redcoats' in Boston in 1770 and the famous 'Boston Tea Party' of 1773. The British Government had attempted to set up a monopoly over the supply of tea to the colonials, and after a stand-off (led by merchants and tea smugglers who faced losing their livelihoods) a cargo ship was boarded in Boston Harbour and its tea, worth £10,000 (around £1 million today) tipped overboard. The War of Independence started in Lexington, Massachusetts on April 19, 1775 and the Declaration of Independence was adopted on July 4, 1776 (though the war itself lasted until September 3, 1783, when it was formally ended in the Treaty of Paris). Although we lost the war, and the country, we left behind the principles of our legal system, democracy and language and played an enormous part in shaping the modern-day USA.

AMNESTY INTERNATIONAL

ENGLISH solicitor **Peter Benenson** founded Amnesty International in London in July 1961 after reading about two Portuguese students who had been sentenced to seven years in jail for drinking a toast to liberty. Portugal at that time was under the rule of the authoritarian dictator António de Oliveira Salazar.

Benenson wrote an article in *The Observer* called 'The Forgotten Prisoners' which described his reaction: 'Open your newspaper any day of the week and you will find a story from somewhere of someone being imprisoned, tortured or executed because his opinions are unacceptable to his government. The newspaper reader feels a sickening sense of impotence. Yet if these feelings of disgust could be united into common action, something effective could be done.'

Since then, Amnesty has spread to more than 150 countries and is now **the world's largest voluntary human rights organisation**, with 2.2 million members and supporters, and its work has led to the release of thousands of political prisoners.

ANAESTHETICS

IMAGINE the scene: it's 1847, and three young doctors are passing round a bottle after dinner. Instead of a vintage port, though, it contains a recently-discovered chemical – chloroform. Each in turn inhales deeply and slips unconscious to the floor. Eventually one of them – **James Young Simpson** – recovers and climbs groggily to his feet. 'My word!' he exclaims, 'This is far better and stronger than ether. It will turn the world upside down!'

Young Simpson and his assistants – Drs George Keith and James Duncan – were not chasing new cheap thrills to compare with opium, but engaged upon a quest which would revolutionise the world of surgery.

At that time, operations were a terrifying prospect. Strapped to a table, given something to bite on and perhaps some alcohol to dull the pain, you would faint from agony if you were lucky, and die of shock if you were not. Medicine had taken a long-awaited leap forward the year before, when ether, another early anaesthetic, had been used for the first time (first in America, and then in Great Britain). But ether had several unpleasant side effects – not least that it was highly flammable and made patients vomit. Young Simpson (1811–1870), a baker's son born in Bathgate, Linlithgowshire, was one of many resourceful Brits striving for something better.

Within four days of discovering the amazing properties of chloroform for himself, he used it to help a woman who was having a prolonged and difficult labour. The first *official* public trial of chloroform in surgery was successfully undertaken by Professor James Miller on 10 November, 1847, in the Edinburgh Royal Infirmary. Operating on a boy aged five, he successfully removed a piece of diseased bone from the child's arm. Miller wrote that there was 'not the slightest evidence of the suffering of pain... He still slept on soundly, and was carried back to his ward in that state. Half an hour afterwards, he was found in bed, like a child newly awakened from a refreshing sleep.' On 20 November, a report was published in *The Lancet* (incidentally, **the world's longest-running medical journal**, founded by Thomas Wakley in October 1823). Despite resistance to chloroform from the church and members of the medical profession, who called it unnatural, it gained the approval of Queen Victoria when the eminent Dr John Snow used it to help her deliver Prince Leopold in 1853 and Princess Beatrice in 1857. Snow, a celibate vegetarian and devout Christian, is regarded as **the world's first anaestheologist**. He experimented on animals to work out accurate dosages and invented equipment to regulate and deliver the amount a patient received.

Strangely, Victoria did not use 'gas and air' – today the world's most common form of pain relief used in childbirth – despite the discovery of its palliative properties more than 50 years earlier. Gas and air is a mixture of nitrous oxide and air or pure oxygen. Nitrous oxide (N_2O) itself had been discovered by the brilliant Yorkshireman **Joseph Priestley** (1733-1804 – see *Fizzy Drinks*) in 1772, but it wasn't until the exuberant chemist and serial inventor **Sir Humphrey Davy** (1778-1829 – see *Electric Light* and *Industrial Revolution*) accidentally alleviated a toothache with it in 1800 that its anaesthetic properties were realised.

Davy, a woodcarver's son from Cornwall with a genius mind and a terror of being buried alive, worked in his early career as a researcher at the Bristol-based 'Medical Pneumatic Institution' – a laboratory and clinic which used human guinea pigs to test the effects of various newly-discovered gases. One experiment carried out there was an ill-fated effort to cure Jessie, the daughter of James Watt (see *Industrial Revolution*), of tuberculosis by making her breathe in carbon dioxide (which actually suffocates). Another attempted TB treatment involved arranging for cows to breathe over patients. It was whilst experimenting with nitrous oxide that Davy 'breathed 16 quarts of the gas in seven minutes' and became 'completely intoxicated'. He said it made him feel 'sublime' and 'superior to other beings'. It also temporarily banished his raging toothache and in his first book, *Researches, Chemical and Philosophical, Chiefly Concerning Nitrous Oxide*, published in 1800, he thus recommended it enthusiastically as an anaesthetic.

Sadly, no-one listened and instead it was nicknamed 'laughing gas' and used as an entertainment at travelling shows. People would pay to inhale a minute's worth and become giddy until the effects wore off. It was not until the 1840s that gas and air was first used in dentistry, and it wasn't used in childbirth until 1881 in St Petersburg.

Between 1933 and 1936, Preston-born anaesthetist **Robert Minnitt** and London engineer **Charles King** created a variety of gas and air machines, for both home and hospital use, by which women in labour could control their own use of the anaesthetic. A further improvement came when in 1961 when **Michael Tunstall**, from Aberdeen, introduced pre-mixed nitrous oxide and oxygen, rather than air, in a single cylinder, replacing Minnitt's machines.

But what of chloroform? Its use was widespread for the next half century or so, but after fears that it was carcinogenic and could cause heart attacks it was abandoned in favour of safer and far more effective anaesthetics (such as halothane, discovered in Britain by **Charles Suckling** at ICI in 1951, and since used around the world). But not before many people had cause to be eternally grateful to James Young Simpson.

ANTISEPTIC

SIMPSON'S pioneering work with chloroform was only half the story: until the middle of the 19th century, patients operated upon successfully *still* stood a 50 percent chance of dying from subsequent infections.

No-one knew why, until a Briton – **Joseph Lister** (1827–1912) – discovered the answer. In doing so, he set in place theatre protocols which have saved millions of lives around the world.

In December 1846, the young Lister – the son of a wealthy Essex wine merchant – watched Robert Liston carry out the first operation in Britain using ether as an anaesthetic (ether having been first used in surgery in Massachusetts by William Morton). The patient, a 36-year-old Harley Street butler called Frederick Churchill, had an infected right knee; Liston took 28 seconds to amputate his leg and the experience made Lister determined to be a surgeon.

Later, while working at Glasgow Infirmary, he was frustrated that many of his patients died despite seemingly effective operations. In 1865, he read a paper by the brilliant Frenchman Louis Pasteur about the germ theory of disease, which suggested that microscopic particles in the air might be responsible for many illnesses. Lister reasoned that, if this was the case, it made sense to attempt to kill the microbes before they reached an open wound. Carbolic acid was used to treat sewage in Carlisle, and he wondered if it might work in a surgical setting.

He carried out **the world's first operation using antiseptic** on August 12, 1865. James Greenlees, an 11-year-old boy, had a broken leg, and the shattered bone had pierced the skin. Lister set the fracture and successfully used carbolic acid to kill any germs surrounding the wound. In 1867, he presented his findings to the British Medical Association, and two years later he described the use of a carbolic spray to the wound and into the atmosphere around an operation. He set up strict antiseptic procedures at his hospital, washing his hands before operations and cleaning instruments and dressings; as a result, the rate of post-operative deaths on his wards fell dramatically.

THE ANYWAYUP CUP

THEY say that necessity is the mother of invention – perhaps they watched toddlers tipping blackcurrant juice all over cream-coloured shag-pile carpets. The horror of seeing a youngster ruin a friend's floor in that way in 1990 inspired British mum-of-three **Mandy Haberman** to invent the 'Anywayup Cup' at her kitchen table.

The world's first non-spill toddler training cup, it contains an air inlet valve which opens when the child sucks on it and seals between sips. Haberman spent several years refining the design until, in 1996, she was happy. Then she filled a cup with blackcurrant juice, wrapped it in white tissue and mailed it to Tesco's buying department, suggesting that if it arrived intact they might like to give her a call. Within a few months, 60,000 cups were being sold per week. Three years later it had risen to seven million a year.

THE AQUARIUM

THE world's first large-scale public aquarium was opened at London Zoo in Regent's Park in 1853. Three hundred different species of marine fish were housed in a huge glass tank, and thousands of gallons of seawater were transported to the exhibition using the newly-developed railway system. People flocked to the new 'Fish House' to gawp at creatures few ordinary members of the public had ever seen before. Scientists were also able to study marine life in something akin to its natural environment. The ambitious project was the brainchild of the English naturalist **Philip Gosse** (1810–1888), who also coined the term 'aquarium' by combining the words 'aquatic' and 'vivarium' (though Latin scholars were sniffy about this, because aquaria were watering places for cattle in Roman times). Gosse's invention led to a craze for aquariums in the home. They became the must-have accessory for prosperous Victorians who wanted to show off: they were the giant-sized, flat-screen, high-definition plasma tellies of their day.

ARCHAEOLOGY

DESPITE being expelled from two convent schools – the second time for deliberately causing an explosion in a chemistry class – **Mary Douglas Leakey** (1913–1996) became a world-renowned archaeologist who made key discoveries that have shaped our understanding of human origins.

Born a spirited only child in London, she developed a passion for archaeology while travelling in France in her youth. In 1948, on Rusinga Island in Lake Victoria in central Africa, she discovered and painstakingly reconstructed the prehistoric jaw and skull of an early primate called *Proconsul africanus*. The fossilised remains, roughly 18 million years old, were the first nearly complete specimen ever found.

In 1959, she pieced together from more than 400 tiny fragments the jaw of the early hominid *Zinjanthropus* (now called *Australopithecus*) *boisei* that she teased from the earth where it had lain buried for almost two million years. This was originally thought to be the 'missing link' between primitive ape-men and early humans, and although this was later disproved it was still a highly important find.

But her most significant discovery came in 1979, in Tanzania, when she discovered the fossilised footprints of early humans. More than 3.5 million years old and preserved in volcanic ash, the lack of accompanying knuckle prints proved that man's ancestors were walking upright much earlier than had previously been believed.

Leakey, who loved Cuban cigars, malt whisky and Dalmatians (she wasn't particularly keen on people, though she had three children), continued to work in the field until she was 70.

ARMED FORCES

FOR good or ill, Britain has been one of the greatest martial nations the world has seen. Our soldiers, sailors and airmen have won many glorious victories, against the odds, all over the world (as well as suffering their share of defeats). Here are snapshots of three great military triumphs – one each from land, sea and air.

Agincourt

THE Hundred Years War between France and England was really a series of on-off battles over the right to the French throne. It actually lasted 116 years, and in the end we lost. But along the way we had some famous victories. Crécy and Poitiers were two of the three greatest, and Agincourt is the third.

The battle happened after a night of teeming rain, in a long and sodden field between two woods outside the village of Agincourt in northern France. It was Friday, October 25, 1415.

English and Welsh soldiers under the command of the Welsh-born **Henry V** were marching to Calais after taking the town of Harfleur (near the Somme, where British troops would die in their tens of thousands 500 years later) when they found their way blocked by a much bigger French army. Exactly how much bigger the enemy force was will never be known: most of our own sources from the period suggest we were outnumbered ten to one, while contemporary French sources suggest the odds may have been closer to five to one. Most modern historians believe that Henry's army numbered around 12,000, while that facing him was 36,000 strong.

The English and Welsh were weary from war, half-starving and dysentery was rife in their ranks. They had marched 260 miles in two-and-a-half weeks. Imagine the dread they must have felt at seeing the fresher, far larger force arrayed against them: accounts from the time talk of men cleansing themselves of their sins before the battle to avoid hell upon death. But before the fighting began, the king made a stirring speech which was later reimagined by William Shakespeare in his play *Henry V*.

Lamenting the overwhelming odds, the Earl of Westmoreland says:

"O that we now had here
But one ten thousand of those men in England
That do no work to-day!"

Henry replies:

"No, my fair cousin;
If we are mark'd to die, we are enow (*enough*)
To do our country loss; and if to live,
The fewer men, the greater share of honour."

He adds:

"We few, we happy few, we band of brothers;
For he to-day that sheds his blood with me
Shall be my brother; be he ne'er so vile,
This day shall gentle his condition;
And gentlemen in England now-a-bed
Shall think themselves accurs'd they were not here…"

This was written around 200 years after the battle, and the play, though based on historical records, is fictionalised. Nevertheless, there is no doubt that Henry spoke to his men, and gave them courage for the fight to come. The French were fresher and better equipped, with perhaps 10,000 knights and men-at-arms (armoured professional soldiers), and 1,200 of them were on warhorses. Henry had 1,000 men-at-arms, 6,000 archers and a rag-bag assortment of a few thousand others.

Many of the French were the sons and grandsons of men who had been killed or captured at Crécy and Poitiers and they were desperate to get at the English and Welsh. If they could close with Henry's men, their superior numbers and armour meant a bloodbath would ensue. And it did, but unfortunately for the French they were on the receiving end.

Henry's longbowmen provoked them into attacking with volleys of arrows at a range of around 300 yards. They raced forwards, but the narrow terrain and pointed stakes driven into the ground funneled them so tightly that they were unable to bring their weapons properly to bear. The charging horses, which wore little or no armour on their sides, were shot down by archers at the flanks and the knights riding them were flung into the mud. There, weighed down by their own armour, they were unable to defend themselves against the nimble footsoldiers who hacked them to death with hatchets or swords. The same fate awaited the French men-at-arms who arrived on foot, as dead, dying and living men were all crushed into a mass of bloody humanity. Many are said to have drowned in the mire.

The hand-to-hand fighting, of a kind we can hardly imagine today, lasted for three hours and left numerous piles of bodies six or seven feet high. By the time the field cleared, 400 English and Welsh soldiers were dead but around 8,000 French had lost their lives (estimates vary). This number was increased when Henry gave the order that all French prisoners be butchered. (At the time, the enemy were mounting a last, half-hearted attack and the king was concerned that he did not have enough men to guard them *and* repel the attack. The following morning, though, he returned to the battlefield and also killed any wounded French who had survived the night in the open.)

After Agincourt, Henry continued on his quest for the French crown. In 1420, the Treaty of Troyes laid down that he would inherit the throne of France on the death of King Charles VI, and he married Charles' daughter Catherine. Unfortunately, both Charles and Henry died two years later, and, though the infant Henry VI became the nominal ruler of both England and France, Charles VII of France declared himself king and later defeated the English (and their French supporters) in battle. England's claim to the French throne was not abandoned until 1802, when we recognised the French Republic.

Trafalgar

TRAFALGAR is among the greatest sea victories in Britain's history (and there have been quite a few). It happened in the middle of the Napoleonic Wars – fought between a series of coalitions of European countries as the French Emperor Napoleon Bonaparte sought to dominate the continent. Napoleon, allied with the Spanish, had for some months intended to invade Britain, to take control of the country's ports and clear the way for French domination of the high seas and the trade that went with it. He could not achieve his ends without destroying the Royal Navy, and Britain could not rest easy while the enemy ships prowled the oceans.

It became clear that a major and decisive battle was needed to settle matters; it happened on October 21, 1805, in the Atlantic some 20 miles off Cape Trafalgar on the Spanish coast.

Napoleon's fleet, under the command of Vice-Admiral Pierre de Villeneuve, was numerically superior, with more, bigger and better-armed ships: some 30,000 men and 2,568 guns faced our 17,000 men and 2,148 guns. But we had **Admiral Lord Nelson**, and it was his tactical brilliance and bravery – and the courage, seamanship and gunnery of our sailors – which won the day.

In the early 19th century, ships tended to sail in strung-out lines. Accepted naval doctrine had it that, to engage the enemy, you sailed alongside and opened fire with a 'broadside'. Cannon fire would then proceed back and forth and, because the sides of the ships were made of thick wood, battles could rage for some time and were often inconclusive.

In order to end Napoleon's ambitions to invade Britain once and for all, Horatio Nelson knew that we needed a conclusive battle – preferably, of course, a win. His stroke of genius was to sail towards Villeneuve's line at right angles. There were pros and cons to this. As he approached, he was very vulnerable to fire from the French and Spanish, and could not return fire – the warships of the day had no cannon on their fore or aft decks. But if he could break through the line, suddenly the guns aboard his ships would be pointing directly along the rear and foredecks of the enemy – where they, in turn, carried no guns and were lightly protected. Thus, his cannon fire would travel uninterrupted and unreturned along the enemy decks.

The two fleets met each other in the morning. At 11.45am, Nelson sent the famous flag signal, 'England expects that every man will do his duty.' (The term 'England' was used at the time to refer to the United Kingdom, and Nelson's fleet included large numbers of men from Scotland, Wales and Ireland.) Then they sailed directly at the enemy in two columns of ships, one under Nelson's command and one under that of Admiral Cuthbert Collingwood. (Just before his column reached the enemy, Collingwood is said to have told his officers, 'Now, gentlemen, let us do something today which the world may talk of hereafter.' It's possible he had this book in mind, but unlikely.)

The fighting was savage. The French and Spanish gunners, many of them actually soldiers unused to firing from moving platforms, engaged the British ships as they advanced. But they were not terribly effective and the British ships got through their lines and were in amongst them. Then our gunners fired. A cannon ball from Nelson's flagship, the *Victory*, could travel the length of a French or Spanish vessel, and take the heads, arms and legs off 30 men or more: within a short time, thousands of allied sailors lay dead or dying, and the battle was on the way to being won. One ship, the *Redoutable*, was left with 99 fit men out of a crew of 643. In these circumstances, there was little the enemy could do but surrender or sail away.

Nelson (1758-1805), a hero at home for his many earlier victories against Napoleon's forces, was renowned for his personal bravery. He always put himself in the heart of the action, and had lost an arm and an eye in previous battles. At 1.30pm, his courage caught up with him when a French musketeer from the *Redoutable* shot him in the shoulder, the ball then lodging in his spine. According to Christopher Hibbert's *Nelson: A Personal History*, he told the *Victory*'s captain, Thomas Hardy, 'Hardy, I do believe they have done it at last... my backbone is shot through.'

He was carried below decks and during one of several visits by the captain to check on him, Nelson did utter the immortal line, 'Kiss me, Hardy', but his final words – shortly after victory was declared at 4.15pm, and the news conveyed to the pale and fading Admiral – were, 'Thank God I have done my duty.'

His body was preserved in a barrel of brandy for the journey home and he was one of 449 British killed. A further 1,241 were wounded, some of whom later succumbed to their injuries, according to Michael Lewis' *The Social History of the Navy 1793-1815*. But on the other side, the death toll was horrific: 4,408 men were killed and 2,545 wounded.

Napoleon's plans to cross the channel were gone forever – 10 years later his dreams of continental domination would be wrecked at the Battle of Waterloo by another great British military leader, the Duke of Wellington – but news of the death of Nelson, *the* hero of his day, was very badly received at home. King George III summed up the national mood: 'We do not know whether we should mourn or rejoice. The country has gained the most splendid and decisive Victory that has ever graced the naval annals of England; but it has been dearly purchased.'

Nelson was buried at St Paul's Cathedral, with thousands lining the streets and an honour guard of 32 admirals and 10,000 troops accompanying his coffin.

The Battle of Britain

JUST as Nelson's victory had prevented Napoleon from invading, so the young pilots who flew in the Battle of Britain almost 150 years later kept another continental megalomaniac – Adolf Hitler – at bay. It was the world's first aerial conflict, fought by desperate and often terrified men thousands of feet above the sleepy English countryside, and it would cost more than 500 Britons their lives.

WWII had begun on September 1, 1939, when Britain declared war on Germany after the Führer's troops invaded Poland. It was little more than a gesture: Hitler was at the head of the most formidable armed forces the world had then seen, and there was nothing, beyond posturing, that we could do to stop him.

In April 1940, German troops invaded Norway and Denmark. Then, on May 10 – the same day that the ineffectual Neville Chamberlain was replaced as Prime Minister by Winston Churchill – Hitler sent his soldiers into France, Belgium and the Netherlands. The Low Countries were overrun in days, France defeated in a fortnight and the British Expeditionary Force – our troops in mainland Europe – was forced to abandon its equipment and retreat to the beaches on the French north coast at Dunkirk. An enormous flotilla of fishing vessels, ferries and civilian boats showed staggering bravery in crossing the Channel to evacuate 338,000 men back to Britain, while the Royal Air Force fought off German bombers overhead.

With America yet to enter the war, and the Soviet Union in temporary alliance with him, Hitler knew that if Britain fell the war would be over almost before it had begun. Had it not been for the English Channel, the RAF and the Royal Navy, he could well have been in Buckingham Palace inside a week. As it was, the Führer ordered a plan to be drawn up for the invasion. 'Operation Sealion' would involve landings on the south coast and airborne troops dropping further inland in September 1940. Crucial to this plan was air superiority: bombers would be needed to sink our destroyers, and this meant our Fighter Command had to be beaten first.

In June 1940, Churchill warned the nation, 'What (was) called the Battle of France is over, the Battle of Britain is about to begin... The whole fury and might of the enemy must very soon be turned on us... Let us therefore brace ourselves to our duty, and so bear ourselves that, if the British Empire and its Commonwealth lasts for a thousand years, men will still say, "This was their finest hour".'

Its exact dates are open to debate, but in the early summer, there was a kind of phoney war, as both sides built up their forces. Britain's factories worked tirelessly to produce Spitfires and Hurricanes – our two fighter planes of the period – while young men were trained to fly them. Despite this, by the time the first German planes appeared over Britain, we were – once again – woefully outnumbered. The

RAF had 640 fighter aircraft, against the Luftwaffe's 2,600 fighters and bombers.

German missions began in earnest on July 10, and from then through to early August they concentrated on probing our air defences, attacking radar stations (see *Radar*) and towns along the south coast. From August 8 until September 6, they turned in earnest on the RAF itself, sending wave after wave of aircraft to attack our airfields, attempting to destroy our planes on the ground or in the air. The most intense fighting came in the middle of August, when the Luftwaffe sent huge formations of aircraft across the Channel to try and break the resilience and resolve of the RAF.

On August 15, according to the excellent Battle of Britain Historical Society website, www.battleofbritain1940.net, the Germans flew 2,000 sorties and lost 75 aircraft, while Fighter Command flew 974 sorties and lost 34. The following day, similar numbers of enemy planes attacked, and **Flight Lieutenant James Nicolson** of 249 Squadron became the only fighter pilot of the war to win a Victoria Cross for his astonishing bravery in shooting down an enemy fighter.

Nicolson was attacked from behind by a twin-engined Messerschmitt Bf 110. Cannon and machine gun fire tore through his Hurricane, ripping open his left leg, leaving him half-blinded from cockpit debris and blood and setting his plane on fire. At this point, most people would have 'bailed out': he was being roasted alive, after all. But as his stricken machine lost speed, the enemy aircraft overshot him and Nicolson, ignoring the flames and smoke which filled his cockpit, dived after it and shot it down. He later said, 'I remember shouting "I'll teach you some manners, you Hun!"' Only as the German crashed below did he think to jump from his own doomed Hurricane. When he landed, he found the heat from the fire had melted the glass in his wristwatch and that he had sustained serious burns to his neck, face, hands and legs. (Sadly, he was killed on May 2, 1945 when his aircraft crashed into the Bay of Bengal. Proving that not *everything* about Britain is great, his elderly widow Muriel was later forced to sell his VC to make ends meet.)

August 18 became known as 'The Hardest Day', when the RAF lost 35 fighters in desperately repulsing hundreds of German attackers; the relentless pace of the Battle was stretching the RAF to breaking point. Its pilots were flying five sorties a day, from airfields which were themselves being bombed and strafed, and were dying in their hundreds. By the end of August, the situation was desperate. In the previous fortnight alone, 295 of our planes had been destroyed and 103 pilots killed; we could not train replacements as fast they were being lost. On the weekend of August 30-31, 65 fighters were destroyed and Hitler was close to achieving his goal.

Then, just as collapse threatened, the Germans fatally switched strategy. On August 24, the Luftwaffe had bombed London by accident (Hitler having given express instructions that the capital was to be left alone, in the hope that we would sue for peace). The RAF replied by bombing Berlin over the following nights, infuriating the Führer. On September 2, he ordered his pilots to target London. This decision gave the RAF a precious few days' respite; aircraft could be repaired, new pilots brought in, radar stations, runways and hangers rebuilt.

On September 7, a series of huge raids involving nearly 400 bombers and more than 600 fighters attacked the docks in the East End of London, day and night, causing terrible damage. But September 15 was to prove the turning point of the Battle of Britain, and the Second World War. The enemy turned their attention back to Fighter Command with a massive raid, involving 1,000 planes, which they thought would wipe out RAF resistance once and for all. Instead, the RAF broke *them*. At one point, every British plane was in the sky with nothing in reserve, but the Luftwaffe's courage and commitment was wavering and the raid was turned back, with 60 Germans shot down. On September 17, Hitler postponed his planned invasion indefinitely. It was never revisited.

Famously, Churchill had said of our young airmen, 'Never in the field of human conflict was so much owed by so many to so few.'

Of 'The Few', 510 were killed. The Luftwaffe lost 3,368 men, killed or captured. It was Hitler's first defeat, and though the war would last almost four more years it was the beginning of the end of him and the Nazis.

ARTISTS

THE world's most famous artists were not British, unfortunately. The Renaissance greats – Leonardo da Vinci, Michelangelo and Raphael – were Italian. The giants of later years, like Rembrandt, Van Gogh or Monet, tended to be French or Dutch. But British artists have given a great deal to the world (and we're not talking about pickled sharks and unmade beds).

Our two most famous painters are **Joseph Turner** and **John Constable**.

Turner (1775-1851) was born in London the son of William, a Covent Garden barber and wig-maker. His mother Mary became insane, and suffered terrible rages, so Joseph spent much of his childhood with relatives. His first job was as an architect's assistant but, aged 14, he was accepted at the Royal Academy of Art. His first exhibit was a watercolour of Lambeth Palace, entered into the Royal Academy's Summer Exhibition in 1790, when he was just 15. His early works depicted coastal scenes and British landscapes, but his later works were influenced by his travels to Europe.

He recorded monumental events – King George IV commissioned a picture of the Battle of Trafalgar in 1822 – and he was called upon by wealthy landowners and industrialists who fancied a painting of their house and gardens. He is remembered as much for his use of colour as his subject matter. The *Oxford Companion to Western Art* says his later works in oil and watercolours show 'unique luminosity' and an 'ability to depict reflections [that] has never been surpassed'. His style influenced many of the later impressionists. He never married and created some 550 oil paintings, more than 1,500 watercolours and 300 sketchbooks of drawings. He bequeathed many of his works to the nation, and London's Tate Gallery now houses the bulk of his work. His beautiful *Giudecca, La Donna della Salute and San Giorgio*, a dreamy, 1841 oil-on-canvas work showing gondoliers in Venice under a blue sky decorated with pink-white clouds, sold in 2006 for £20.5 million. In doing so, Turner almost doubled the previous record for a British picture – that was *The Lock* by John Constable, which sold for £12 million in 1990.

Constable (1776-1837) was Turner's great contemporary. Born the son of a wealthy corn merchant in East Bergholt, Suffolk, he worked in the family mills after leaving school, but later persuaded his father to allow him to enrol at the Royal Academy. In 1816, he married a childhood friend, Maria Bicknell. To make ends meet he sold portraits, but his passion lay in the countryside – he sought particularly to reflect the changing light and patterns of the clouds moving across the sky.

His most famous painting, *The Haywain*, was completed in 1821, and shows a cart in the River Stour in front of Flatford Mill, the watermill owned by his father. It caused a sensation when it was exhibited in France in 1824, winning a gold medal at the Académie des Beaux-Arts in Paris, the leading art event in the world at the time. Indeed, Constable's work was far more popular in France than it was here – he sold only 20 paintings in Britain during his lifetime, whereas he was revered across the Channel for his use of colour. But he refused to move there, saying he would rather be poor in England than a rich man abroad. *The Haywain* can be viewed at the National Gallery, London.

ARSENIC

BEFORE English chemist **James Marsh** (1794-1846) came up with an accurate test for arsenic poisoning in 1836, it was commonly referred to as 'Inheritance Powder'. Odourless, tasteless, readily available and easily slipped into food or drink, a potential murderer could quickly dispose of an elderly relative and claim a legacy. For instance, serial killer Mary Ann Cotton, from Durham, is believed to have murdered 24 people with arsenic, including nine of her children, her mother, her stepchildren and several husbands. She collected life insurance payouts after each death and was eventually hanged for murder in 1873. Arsenic poisoning also mimics recognised illnesses – Cotton's victims' deaths were put down to gastric fever – making it virtually undetectable.

Marsh, from Woolwich, London, gave evidence in a murder trial in November 1833. John Bodle, from Plumstead, had been accused of putting arsenic in his grandfather George's coffee. Marsh used an existing test to detect arsenic in the cup but the evidence deteriorated over time and his forensic proof was dismissed. Bodle was acquitted, only to admit the crime later. At that time, you could not be tried twice for murder so he escaped justice; outraged at this, Marsh spent years working on scientific experiments before coming up with what is now known as the Marsh Test. He put arsenic-tainted material in a closed bottle with sulphuric acid and zinc. This created a gas which escaped through a narrow tube in the lid of the bottle. If the gas left a silvery-black powder behind when ignited, it proved the existence of arsenic. The Marsh Test became the standard proof for arsenic poisoning.

ASPIRIN

WHILST strolling through an Oxfordshire meadow in 1757, an English vicar called Edward Stone idly pulled a piece of bark off a willow tree and nibbled on it.

Its surprising bitterness reminded him of the taste of the malaria treatment, quinine – itself derived from a tree, the South American cinchona. Wondering whether willow might also have therapeutic properties, he made a dried powder from the bark and began taking it to treat his chills and fevers. Over the following five years, he also gave it to 50 members of his parish in Chipping Norton. It proved an effective remedy for many of their minor ailments. Stone (1702-1768) had discovered **the world's best-selling painkiller**. In 1763, he wrote to the Royal Society with his findings. The Society's website describes this as the 'first research into salicylic acid, the active material in aspirin'.

Aspirin in tablet form came much later, after German scientist Felix Hoffman synthesised the acid in 1897. It was soon used all over the globe, though scientists had no idea how it worked. In 1966 the *New York Times* called it 'the wonder drug that no-one understands', and it took another Englishman, pharmacologist **John Vane**, to discover why it was so effective. He found that aspirin works by blocking production of a hormone-like substance called prostaglandin, which triggers pain, fever and inflammation. He also found that even a very low dose of aspirin stops the production of a prostaglandin called thromboxane, which plugs tears in blood vessels, making the drug useful for stopping the formation of blood clots and helping to prevent heart attacks and strokes. These discoveries earned Vane (1927-2004), from Wythall, Worcestershire, a knighthood, more than 50 honorary degrees and fellowships and a Nobel Prize. Today, some 50 billion aspirins are taken annually around the world.

BABY BUGGIES

A TEST pilot and aeronautical engineer changed the lives of millions of parents when he made **the world's first baby buggy** – a lightweight collapsible pushchair. After watching his daughter struggle with a cumbersome pram, **Owen Maclaren** set to work in the stables of his farmhouse in Barby, Northamptonshire.

Maclaren, who was born in Saffron Walden, Essex, in 1907, had designed the Spitfire undercarriage, so was familiar with lightweight, durable materials that could be made to fold neatly. He patented the Maclaren baby buggy in 1965 and production started in 1967. The first model had a frame of tubular aluminum and weighed just 6lb – lighter than the average new-born – and folded down to the size of an umbrella. Within a few years production was up to 600,000 a year.

BADMINTON

IF you suggest that any given sport is 'British' in origin, someone, somewhere, will claim that cave drawings or ancient scrolls prove that you're wrong, and that something vaguely similar was first played, in some form, at some time, by the Persians or the Incas or the Romans or A N Other ancient civilisation. These claims come with varying degrees of evidence, and it would be foolish to disregard them all. However, what is undeniable is that the vast majority of world sports – as they are recognised and played today – sprang from these islands. Our love of order, genius for codification and domination of much of the world during the 18th and 19th centuries enabled us to bring together disparate elements of various pastimes and to standardise their rules.

This should be a great source of pride – unlike our subsequent performances in the sports we have created, sadly.

Badminton is one of these games (others will follow, in alphabetical order). Something a *bit* like it *might* have been played by the Chinese in the 5th century, or in 16th century Japan, or in 19th century India, but badminton, as found on squeaky wooden floors in freezing sports halls across the length and breadth of this nation (and elsewhere), is indubitably British.

It's named after Badminton House in Gloucestershire, where it was first played officially in 1873 at a party held by the **Duke of Beaufort**. The first shuttlecocks were champagne corks with feathers attached and it was originally called 'The Badminton Game'. The International Badminton Federation now has around 140 member nations and estimates that 200 million people play worldwide. The best feathers for a shuttlecock apparently come from a goose's left wing (seriously) and it is the fastest projectile of all racquet sports, having been timed at 206mph.

BASEBALL

THE USA's national sport – baseball – is actually British.

The game's folklore has it that it was created by a General Abner Graves in Cooperstown, New York State, in 1839. But we Brits were probably quite bored with the game by then – we'd been playing it for almost a century, after all. A diary kept by the Surrey historian William Bray documents a game in Guildford in 1755, Jane Austen mentions the sport in *Northanger Abbey*, written in 1797-8, and a German book from 1796 features seven pages on the rules of 'Englischer Baseball'.

Informing Americans of this tends to bring on paroxysms of rage – the effect is doubled if you refer to your interlocutor as 'old boy' or 'my dear old thing', and smile condescendingly while imparting the information.

To make matters worse (or better, depending on your perspective), we can claim basketball, too. It was invented by James Naismith in Massachusetts in 1891, but Naismith was born in Ontario in 1861 and at that time Ontario was a British North American Colony. Thus, technically, Naismith might be said to be British. This is slightly tenuous, but it is also good for annoying our transatlantic friends. (If you're in a *really* mischievous mood, point out that American football is really just rugby with helmets, forward passes and interminable ad breaks.)

BATHTUBS

THE enamel bath is one of the world's simplest luxuries. A Briton, **David Buick**, is behind it.

Buick was born in Arbroath, Scotland in 1854, but later emigrated to Detroit, Michigan. He went to work for a plumbing firm, where he invented the process of permanently bonding porcelain enamel to cast iron baths which is still used to this day. He also founded Buick Cars in 1902 but sadly, he wasn't much of a businessman; the company veered from one financial disaster to another and he was forced to sell out to a business partner in 1906. Had he been able to hold on to his shares they would have been worth $10 million by 1921, but he died, impoverished, in 1929, without enough money to buy one of the cars that bears his family crest.

BALL BEARINGS

WELSH ironmaster **Philip Vaughn** took out the **first patent on ball bearings** in 1794. At that time, machinery wore out quickly as moving parts rubbed against each other and Vaughn, from Carmarthen, came up with the idea of using solid balls in a groove to take the strain and reduce friction. His patent was for a ball bearing system to support the axle on a carriage. His invention is now used in everything from roller skates and car wheels to electric motors and computer disk drives.

BEER

ALMOST every country has a fizzy, yellowy drink called 'lager' – Britain's richer, darker beers, undoubtedly the world's finest, set us apart from the common herd.

Home-brewed beer was our main drink in the Middle Ages – the boiling of the water in the brewing process, and its alcoholic content, had a sterilising effect which made it safer to drink than water. By the time of Henry VII (1485-1509), ladies-in-waiting were allowed a gallon of beer for breakfast – a royal tradition Prince Harry is believed to maintain to this day.

In the centuries before modern quality control, beer went off quickly. **Dr Alexander Nowell**, Dean of St Paul's Cathedral in London from 1560 to 1602, discovered by accident that beer stored in corked bottles stayed fresh. He had taken some ale on a fishing trip, but left it behind in a river. When he found it a few years later, the cork came away with a pop and the taste and quality of the beer was still good.

Britain's oldest existing brewery is Shepherd Neame, founded in 1698 by Captain Richard Marsh, Mayor of Faversham in Kent. But the brewing capital of Britain is Burton on Trent. Benedictine Monks started brewing at Burton Abbey in 1002, but the arrival of draught Bass, first produced in the town in 1777, was the significant moment. Bass was drunk aboard The Titanic – not, as far as we know, by the Captain or lookouts – and its distinctive Red Triangle is Britain's oldest trademark. It features in 40 of Picasso's paintings, the impressionist Edouard Manet's 1882 word *Bar at the Folies-Bergère* and Quentin Blake's illustrations for Roald Dahl's *The Twits*.

The great British country pub, often with thatched roof, roaring fire and 'friendly' locals, is one of the great modern wonders of the world.

BETA BLOCKERS

TWO drugs developed by a Scottish pharmacologist have transformed the lives of millions of people worldwide.

Sir James Black, the fourth son of a Fife colliery manager, was born in 1924 and trained as a doctor at St Andrew's University. Preferring research to patients, he developed **the world's first 'beta blocker'**, Propranolol, in 1964. Beta blockers are used to treat angina, high blood pressure and heart failure, and work by inhibiting the effect of adrenaline on the heart muscle, effectively allowing it to relax. It is regarded as among the greatest medical discoveries of the 20th century.

Black's second breakthrough came in the early 1970s when he came up with an anti-ulcer drug which works by blocking the absorption of histamine, which irritates the stomach lining. It drastically reduced the number of operations on people with stomach ulcers. In 1988, he was awarded the Nobel Prize for Medicine for both drugs. He is still active, and currently working on a range of new ideas including a cancer-blocking drug.

THE BICYCLE

IN 1817, the German Baron Karl Wilhelm Friedrich Christian Ludwig Freiherr Drais von Sauerbronn invented the 'Laufmaschine'. This two-wheeled frame with a seat upon which the rider sat, pushing himself along with his feet, spread to London during the summer of 1819. There was a brief a craze for them, which ended when careless riders started being fined.

Kirkpatrick Macmillan, born the son of a Dumfriesshire blacksmith in 1812, is said to have built a bicycle some time around 1839 – it used pedals on the front wheel to drive connecting rods which turned the rear wheel. It was extremely heavy and riding it was very hard work, though in June 1842 Macmillan rode 68 miles to Glasgow on it. According to www.undiscoveredscotland.co.uk, the trip took him two days and led to him being fined five shillings for knocking down a small girl in the Gorbals. He never patented the invention.

In the 1870s the Penny Farthing, invented by **James Starley** in Coventry, was born. This famously had an enormous front wheel and a much smaller rear one: with the pedals still attached to the front wheel, its greater circumference allowed for higher speed.

But there were obvious issues associated with riding through the cobbled streets of the day on an inherently unstable vehicle with no brakes – chiefly the possibility of taking 'a header', particularly when one's head was positioned 12ft from the ground. (Some Penny Farthing riders hooked their feet around the handlebars when descending hills at speed, so that if disaster struck at least they would be catapulted feet-first.) So it was with some relief that cyclists greeted the arrival of **the world's first modern bike**, the 'Rover Safety Bicycle', in 1885. Designed in Coventry by **John Starley**, James' nephew, it had two wheels of similar size, a pedalled chain and a saddle and handlebars where logic suggested they should be. It was a huge success, improved further by Scotsman **John Boyd Dunlop**'s invention of the pneumatic bicycle tyre in 1887, and was exported all over the world. ('Rover' is still the Polish word for bicycle.) Rover later branched into motorcycles and cars, and no longer exists as a British company.

BIOMETRICS

IN the modern world, ID is a major issue. With fake passports and bogus documents floating around in their thousands, wouldn't it be useful if people could be identified beyond all doubt? Thanks to **technology pioneered by a British computer expert**, they now can.

As long ago as 1949, the Cambridge ophthalmologist (and Bloomsbury Set member) **James Doggart** (1900-1989) had suggested that iris recognition might offer a way of establishing identity. The patterns in the coloured iris of the human eye are unique to each individual. Unlike fingerprints (the earliest biometric ID system), which can be damaged or even removed deliberately with chemicals, they never change. Unlike DNA, even identical twins have different eyes. Thus, if you can scan a man's iris – and his iris is on your database – you can identify him. The problem was, no-one could work out how to turn this science fiction idea into a reality – the patterns on the human iris are *so* random that computers couldn't recognise them.

Cambridge University professor **John Daugman** changed that when he developed IrisCode – a complex piece of software containing an algorithm which allows the eyes to be scanned and read in a foolproof way.

What does this mean in practical terms? The United Arab Emirates installed Daugman's technology as part of its border controls in 2001. Since then, it has scanned more than a trillion irises without a single false match. Along the way, the authorities have caught more than 300,000 people trying to re-enter the United Arab Emirates with dodgy travel documents. In the UK, the Iris Recognition Immigration System allows passengers registered on a database to enter the UK without a passport at Heathrow, Gatwick, Manchester and Birmingham airports. A special camera captures an image of your eye, this is transformed into digital code and then compared with those stored on the database at a rate of a million cross checks per second. A match is usually made within 10 seconds – as opposed to the half hour or more it can take to travel through passport control. In the future, it's anticipated that iris recognition systems will spread to areas like cash cards – unlike a PIN, you can't forget your iris.

BLACKPOOL ROCK

SEASIDE rock was developed for Victorian factory workers to take home as a cheap and cheerful gift after a seaside day trip. The first lettered rock was probably made by a former Burnley miner, **Ben Bullock**, in 1887. He was a sugar boiler with his own factory in Dewsbury, and sold his sweets at local markets. It is believed that the first words he put in a stick of rock were *Whao Emma* – a well-known music hall song of the time – but after a holiday in Blackpool, he made the first Blackpool rock and sent samples to the resort. As word spread, he became inundated with orders from seaside resorts throughout the country.

BLINDNESS

A CURE for the most common form of blindness, which affects 14 million people in Europe alone, has been developed by British scientists.

Using stem cell technology (see *Stem Cells*), experts at London's Moorfields Eye Hospital and the Institute of Ophthalmology at University College London have devised a treatment for age-related macular degeneration (AMD). In this condition, cells under the retina degrade, and this cell loss stops light receptors from working. Embryonic stem cells – which have the ability to develop into all types of body tissue – are grown into replacement cells in the lab, placed on an artificial membrane and implanted in the eye. Clinical trials in rats and pigs have been successful and human tests are planned.

It could be a routine procedure by 2015. In a *Sunday Times* interview, **Professor Pete Coffey**, who is leading the research, said it would take 'less than an hour… it really could be considered as an outpatient procedure. There are about half a million people in the UK at the moment who would qualify for this treatment – about 25 per cent of the population aged over 65.'

In a separate project at Moorfields, the world's largest specialist eye hospital, sufferers from a different condition have also had their sight restored. A group of patients, either chemical burn victims or sufferers of a rare genetic disease known as aniridia, had implants of new limbal cells (cells under the eyelid which maintain the transparent layer on the outside of the cornea). **Dr Julie Daniels**, heading the research team, told the *Daily Telegraph*: 'We can't restore sight completely yet as the material we are growing the stem cells on is slightly opaque, but before the surgery the patients were barely able to recognise when someone was waving a hand in front of their face… we have restored their vision to the point they can read three to four lines down the eye chart.'

(See also *Cataracts*.)

BLOOD

WILLIAM Harvey (1578-1657), a farmer's son from Folkestone in Kent, made a huge contribution to medicine with his **discovery of the workings of the heart**.

It may seem obvious to us, but 400 years ago very little was known about the workings of human organs. At that time, it was believed that the liver converted food into blood, that the lungs were responsible for moving it around the body and that various other parts consumed it in order to maintain life and energy. The function of the heart was unknown.

Harvey qualified as a doctor in 1604 and married Elizabeth Browne in the same year. He started work at St Bartholomew's Hospital, where he carried out experiments on live animals. He punctured a dog's artery and watched blood spurting with the beating of the heart; from this he concluded that the heart was a pump. He dissected the corpses of executed criminals to learn more about the circulatory system (for instance, he realised that the valves in veins prevented blood flowing the wrong way) and in 1628 he published a book, *Exercitatio anatomica de motu cordis et sanguinis in animalibus* (An Anatomical Study of the Motion of the Heart and of the Blood in Animals), which outlined his discoveries. It was controversial – other scholars didn't believe his theory – but it became the basis for modern research on cardiovascular medicine.

During his lifetime, Harvey was physician to King James I and King Charles I. The son of Charles' friend Viscount Montgomery had fallen from his horse as a boy and a resulting abscess had left him with a permanent hole in his chest. He covered the hole with a removable metal plate, and Harvey used him to demonstrate the beating heart. Harvey wrote that he was able to insert 'three fore-fingers, and my thumb' and 'took notice of the motion of his heart'. What young Montgomery thought of all this is unfortunately not recorded.

Harvey also investigated reproduction in mammals, and was the first to propose that humans reproduced via the fertilisation of an egg by sperm. Despite this, he had no children – possibly because his wife kept a parrot that nestled in her bosom at night, which would surely kill the passion of a man not already brought low by gout and kidney stones.

BLOOD TRANSFUSIONS

THE world's first successful blood transfusion saved the life of a young mother in Britain in the early 19th century. Since then, the procedure has saved millions more lives.

Transfusions had been attempted before. In 1492, Pope Innocent VIII died after he was given blood from young boys (the blood was administered orally which can't have helped). In 1667, a youth in France supposedly recovered after being given sheep's blood intravenously by

Dr Jean-Baptiste Denis, physician to King Louis XIV. This does not square with modern understanding of the science, however, and it is likely that the quantity transfused was tiny, if it even happened.

Animal-to-animal transfusions took place in Britain in the 1600s – the Wiltshire-born clergyman Francis Potter, a Cornish doctor called Richard Lower and the great architect Sir Christopher Wren all made progress in this area. But the first definitive blood transfusion, using human blood and saving the life of someone who would otherwise have died, was carried out by the obstetrician **James Blundell** in 1818*.

Death in childbirth from post-partum haemorrhaging was relatively common, and Blundell, a sensitive man, found his inability to save the lives of his patients distressing. He carried out experiments on animals using home-made utensils and finally applied his method to a young woman, extracting blood from her husband's arm and transferring it to her. In later years, he performed 10 further transfusions at Guy's Hospital, London, and five of them were beneficial. (The fact that five were not reflected the failure to understand that blood comes in four different groups, and that these can be incompatible – this was not discovered until the turn of the 20th century, by Dr Karl Landsteiner of Vienna. Landsteiner received the Nobel Prize for his discovery, though this didn't stop him dying of a blood clot.)

Blundell (1790-1878) did identify the clotting that occurred when transfusion was delayed, however, and he developed an elaborate system with a funnel and syringe, the Gravitator, to overcome this problem. In 1840 in London he performed **the first successful whole blood transfusion** to treat haemophilia. He died a very rich man, worth some £30 million in today's terms, as his methods and equipment were taken up around the world.

---

* At least, according to the NHS National Blood Service – other sources have him carrying out the first successful transfusion in 1828, and say that it was the *death* of the unfortunate woman in 1818 that inspired him to increase his efforts on transfusion.

BOND, JAMES BOND

HE'S the world's most famous secret agent and, naturally, he's one of ours (though his mother was Swiss).

James Bond, a Commander in the Secret Intelligence Service (MI6), has saved the world from destruction many times. Along the way, he has outwitted dozens of sinister, unhinged enemies – people like the golden-gun-toting assassin Francisco Scaramanga, the sadistic torturer Rosa Klebb and the mad scientist Dr Julius No. Perhaps his most persistent foe was the evil genius Ernst Stavro Blofeld, head of the global criminal organisation SPECTRE. Blofeld came close to global domination on a number of occasions, but the British spy always managed to thwart his plans. Eventually, 007 killed Blofeld, though how he did this depends on whether you are reading one of British author Ian Fleming's original books, or watching a Bond film (in the 1964 novel *You Only Live Twice*, Bond strangles Blofeld, whereas in the 1981 movie *For Your Eyes Only* the villain dies after Bond tips him down a factory chimney stack, in his wheelchair, from a helicopter, after a series of unlikely manoeuvres).

Sadly, the star of *Live And Let Die*, *Diamonds Are Forever* and *From Russia With Love* is fictional. But he is based on real spies and real events in Fleming's own career in Naval Intelligence during WWII.

Our secret services are some of the best in their field. Among the most interesting was the Special Operations Executive (SOE). Set up in July 1940 to carry out spying, sabotage and assassination missions against the Germans, SOE agents undertook operations around the world, using mini-submarines and silenced pistols, and carrying forged identity documents, ration cards and passports from dozens of different countries. (SOE fakers made clothes and even cigarettes designed to look exactly like those found in Germany or other countries where agents were headed). Other SOE inventions included tins of oil containing abrasive additives which could be swapped for real ones in trains or factories, hollowed-out lumps of coal filled with explosives to be left next to a ship's boiler, landmines disguised as cow pats and suicide pills made to look like buttons.

Many of their missions were deadly serious and extremely dangerous. Women SOE agents worked mainly in France; some undertook the particularly hazardous job of chatting up enemy officers in bars, luring them to a back alley and then killing them. Out of 39 female operatives sent across the Channel, 13 were caught and executed.

The Secret Intelligence Service was set up in 1909 with the job of countering foreign espionage in Britain and doing some of our own abroad. It was split into two sections, one specialising in activity at home (Military Intelligence 5, or MI5) and MI6 working overseas. Among *their* most inventive WWII schemes was one which involved manufacturing a special edition 'Monopoly' game to go in Red Cross parcels destined for British POWs held in Germany. The Germans were happy to accept these games as they kept the prisoners occupied, but they might have been less keen had they known that the games contained more than the usual Scotty dog and car and were, in fact, one big 'Get Out Of Jail Free' card. The SIS Monopoly sets included additional playing pieces like metal files and compasses, real money and silk maps featuring local safe houses sewn into the box lids.

THE BOUNCING BOMB

BARNES Wallis possessed one of the most inventive minds of the Second World War. A scholarly and bespectacled man from Ripley in Derbyshire, he is best known as the inventor of the 'Bouncing Bomb' which was used to destroy two huge dams believed to be vital to German industrial production in the Ruhr Valley. The dams were protected by torpedo netting, so Wallis devised a cylindrical munition which would skip across the water – and the nets – and sink against the dam wall before exploding. To succeed, the bombs had to be released at precisely the right height, speed, position and heading, which meant the Lancaster bombers delivering them had to be flown straight and level into a virtual wall of incoming German anti-aircraft fire. It was an almost suicidally brave undertaking and eight out of the 19 aircraft involved were lost. However, two of the three dams targeted were destroyed, and the German war effort was interrupted, albeit temporarily. To this day, 617 Squadron are known as the 'Dambusters'.

Wallis (1887-1979) also came up with the first 'Bunker Busting' bombs. Reasoning that the force of an explosion would be more potent if it occurred underground (and so was concentrated by the earth instead of dissipating), he designed a bomb with a hardened tip which would hit the ground at supersonic speeds and penetrate to a depth of several metres before detonating. These were used against German V1 rocket launch sites, submarine pens and the *Tirpitz* battleship.

After the war, he pioneered swing wing aircraft design (though much of his work was passed to the Americans who used it to develop the F-111 fighter-bomber) and worked on plans for a hypersonic aircraft (five times the speed of sound, around 3,500mph). This was decades ahead of its time, and still poses the biggest unsolved problem in aviation today. In 1951, he was awarded £10,000 from the Royal Commission on Awards to Inventors (£700,000 today) – he gave it to a school for the children of RAF personnel.

BOXING

MOST cultures have had martial sports – the ancient Greeks, in particular, fought with gloved fists – and few of the truly great boxers have been British. The legendary names tend to be American, or Mexican, or Cuban. But Muhammad Ali, Julio Cesar Chavez and Kid Chocolate all fought under rules created by a Welshman, **John Chambers**, in 1865. These were formally endorsed by the 9th **Marquess of Queensberry** in 1867, and these 'Queensberry Rules' have, with occasional adaptations, formed the basis for boxing ever since.

CARRIERS

THE world's first aircraft carrier was built in Scotland in 1918. *HMS Argus* was the brainchild of **James Graham**, 6th Duke of Montrose, who was born in London in 1878. Glasgow shipbuilding firm William Beardmore was working on an Italian passenger liner called *Conte Rossa* when WWI broke out in 1914. Orders were given to convert her into a warship with a full-length runway to allow planes to take off and land, and the Duke, a marine engineer, oversaw the project. Weighing in at 14,550 tons, *HMS Argus* was completed just weeks before the end of the war, at a cost of £1.3million. She was slow and small and saw little action, but she was used as a template for aircraft carrier design from then on. (Graham also took first moving photographic record of a total solar eclipse, in 1899 in India.)

THE CASH MACHINE

ASK who invented the hole-in-the-wall cash machine, and you'll get a barrage of different replies. Some say it was Luther George Simjian, a Turk who patented a never-used prototype in 1939. The Americans argue it was created by their own Don Wetzel, who began work on his concept in 1968 after getting tired of waiting in a bank queue. But the real battle for the credit is between two Scots.

The first is retired engineer **John Shepherd-Barron**, from Tain, Rosshire. In the best traditions of inventors since Archimedes, he was lying in the bath, mulling over a way to get cash from his bank after closing time (it was the 1960s, and banks were open for about three hours a day back then), when inspiration arrived. 'I hit upon the idea of a chocolate bar dispenser, but replacing chocolate with cash,' he told the BBC. His machine used special cheques coated in radioactive Carbon 14, rather than plastic cards. The cash machine would recognise the cheque, match it to the customer using a personal identification number, and issue £10 (£130 today).

Using his technology, the world's first cash machine opened outside a branch of Barclays in Enfield, London, in 1967. The first customer was Reg Varney, an actor who starred in the truly dreadful 1960s sit-com *On The Buses*. Mr Shepherd-Baron answered fears about the use of radioactivity by stating that 'you would have to eat 136,000 such cheques for it to have any effect on you'. There is no record of anyone trying this to find out.

The second Scotsman to make the claim is **James Goodfellow**, from Paisley in Renfrewshire, also a retired engineer, who came up with **the world's first modern cash machine** using plastic cards with magnetic strips allied to four-digit PIN numbers. He patented his idea in 1966 – a year before his rival's cash dispenser was installed, and two years before Wetzel applied for *his* patent in the USA. In 2005, Mr Goodfellow was disappointed when newspapers were filled with reports that Shepherd-Baron had just been made an OBE for services to banking and was the inventor of the ATM. Goodfellow told *The Guardian*: 'For him to go down in history as the inventor of the ATM really stuck in my throat… He invented a radioactive device to withdraw money. I invented an automated system with an encrypted card and a PIN number, and that's the one that is used around the world today.' He received his own OBE the following year.

CATARACTS

IN 1949, the ophthalmologist **Harold Ridley** implanted **the first artificial lens** in a cataract patient – the first step on a road which improved the lives of millions of people around the world with failing eyesight. During WWII, Ridley – who was born in Kibworth Harcourt, Leics, in 1906, and died 'Sir Harold' in Salisbury, Wilts in 2001 – treated allied fighter pilots whose eyes were full of Perspex after their cockpit canopies had been blown apart by gunfire. He noticed that the Perspex splinters were not rejected by the men's immune systems, and realised the applications that the material might have.

CATSEYES

WHEN **Percy Shaw** died in 1976, untold millions of his 'self-illuminating road safety devices' − or Catseyes − had been produced, and thousands of road accidents had been averted.

But the inventor from Halifax was not the multi-millionaire he should have been: he died in the house in which he had grown up, and left an estate worth a relatively modest £193,500 (a little more than £1.5 million today).

Shaw was born in 1890, 20 years before mass-produced cars even existed, the son of a textile factory worker and one of 14 children. His eureka moment came as he was driving home from his local pub along a dangerous stretch of road in thick fog. Unable to see more than a couple of feet ahead, he pondered the need for a cheap way of marking out the road (legend has it that the method appeared in front of him when his car headlamps were reflected in the eyes of a cat in the grass verge).

He spent years developing his idea while working as a road repairer: he knew his device had to be bright enough to light up roads at night, to work in all weathers, to be strong enough to suffer lorries driving over it and to require minimal maintenance. In 1934, he finally patented the Catseye road stud. It was made up of two bullet-shaped pieces of glass, housed in a rubber pad and mounted on metal, and was even self-cleaning − when a vehicle drove over it, the rubber coating would push up against the reflectors and wipe them clean.

A year later he set up Reflecting Roadstuds Ltd and opened a factory in Boothtown. In 1936 he carried out an experiment, placing 50 Catseyes on a dangerous stretch of road near Bradford. The number of accidents dropped and drivers said it was due to his invention. In 1937, he won a competition staged by the Department for Transport which was looking for a way to cut road accidents. By 1965, Catseyes were being sold all over the world and Shaw received an OBE for services to exports.

He never married, and ploughed much of the money he made back into his company. His only luxuries were a Rolls Royce, cans of (virtually undrinkable) Worthington's White Shield and the odd round of golf. Oh, and four television sets which he kept on at all times (one permanently tuned to BBC1, one to BBC2 and the third to ITV, and a spare in case one broke).

His invention has since been upgraded by another Yorkshireman – **Martin Dicks**, a former fireman from Doncaster – who invented a solar-powered version of the Catseye. Called The SolarLite, it offers 10 times the visibility of traditional Catseyes and lasts far longer. His improved version has been approved by the Department of Transport and is now being installed in British roads.

CHAMPAGNE

'COME quickly, I am tasting stars!' These are supposed to have been the words of the French monk Dom Perignon at the moment he discovered champagne. He could have tasted them decades earlier if he'd crossed the channel and visited the British scientist **Christopher Merrett**.

In 1662, Oxford graduate Merrett (1614-1695), from Winchcombe, Gloucestershire, presented a paper to the Royal Society describing how adding sugar or molasses to wine caused secondary fermentation in the bottle and so created sparkling wine. So, *pardon messieurs*, the 'quintessentially French' drink is actually a *British* invention.

The earliest French document to mention sparkling champagne was written in 1718, and it referred to experiments with secondary fermentation carried out 20 years before. The first French champagne house, Ruinart, was established in 1729, fully 67 years after Merrett first published his discovery.

Of course, once you've invented champagne, you're faced with the problem of storage. The average bottle of champagne contains 50 million bubbles at three times the pressure of a car tyre; ordinary wine bottles tend to explode under that kind of stress. So you need special toughened glass for the job – and who came up with *that* solution? Yes, that was us, too.

Admiral Robert Mansell (1571-1652) set up a glass factory in Newcastle after retiring as a naval officer. He discovered that by adding iron and magnesium to the molten mixture in his furnace he could make much stronger glass. It was this 'verre Anglais', and its successor materials, which the French pinched for their fizz. Oh, and we also began employing corks to stop up bottles while they were still using wooden bungs wrapped in hemp, which did not produce a very good seal.

It would be crazy – even in a book celebrating Britain's greatness – to suggest that our viticulture is better than that found across the Channel. French champagne is the acme of sparkling wines, but UK winemakers like the award-winning Ridgeview Estate in Sussex are proving that Merrett's methods work well here, too.

THE CHANNEL TUNNEL

IN 1994, the Queen and the then French President François Mitterrand opened the Channel Tunnel. For the first time in 8,500 years, after meltwater from Mesolithic glaciers had helped to form the English Channel and cut Britain off from the continent, we had a physical link to Europe.

The American Association of Civil Engineers promptly named it **one of the Seven Wonders of the Modern World**, and their European counterparts acclaimed it as 'the greatest infrastructure success of the 20th century'. (It's not an entirely British achievement – the French played a significant role in it, but they can discuss their contribution in a book called *So That's Why They Call It Great France*.)

The idea of connecting these islands to the European mainland was first mooted in 1751, but it was not until 1802 that a French engineer, Albert Mathieu, came up with anything like a feasible plan – he suggested a tunnel lit by lamps, ventilated by huge chimneys and broken by an artificial island halfway across where carriage horses could be rested.

Over the next century a number of other proposals were put forward, but the ongoing wars, disputes and petty squabbles between the British and the French meant all came to naught, though the sheer scale of the eventual project (it took 15,000 workers and 170 million man-hours, and cost £9 billion) suggest that it would have been far beyond the capabilities of 19th century engineers and nations anyway.

Among the more serious proposals was that of another Frenchman, Aime' Thome' de Gamond, in 1856. He amplified Mathieu's design, adding a dozen additional islands and went so far as to consult **Isambard Kingdom Brunel** (who had tunnelled under the Thames – see *Industrial Revolution*) over technical matters. His plan was favourably received by both Napoleon III and Queen Victoria, but never progressed beyond the drawing board.

Later in the century, the Welsh civil engineer William Low (he was born in Scotland, but lived and worked in Wrexham) suggested a pair of parallel, single-track rail tunnels, connected by cross passages – not a dissimilar design from that which now exists. Thousands of geological samples were taken and digging actually began, but when the tunnel reached 1,000 yards or so work was halted because of fears that it might be used against us. General Sir Garnet Wolseley, 'The Great Anti-Tunneller', claimed that 20,000 men could invade England via the Tunnel in four hours, and that was enough to end the scheme. (This fear persists; in the 1970s, when 'Le Tunnel Sous La Manche' was back on the agenda, the Ministry of Defence suggested installing a nuclear device in the as-yet unbuilt link in case conventional explosives were not enough to destroy it in time of national emergency. As one official pointed out, this was probably not worth worrying about. 'Frankly,' he wrote in a briefing note, 'if we ever reach a situation where Warsaw Pact conventional forces reach Calais without a strategic nuclear exchange having occurred, then I think the Channel Tunnel will be an irrelevance.')

Eventually, a conglomerate of British and French companies started work in 1987.

The first boring machine (TBM) began excavation on December 1 at Folkestone on the UK side, aiming to meet a French TBM coming from Calais. Both were laser-guided: the TBMs could not reverse (they were eventually either dismantled and carried back out of the tunnels or entombed in concrete chambers under the seabed) and if they had missed each other by more than 8ft the resultant 'kink' in the finished tunnel would have been structurally disastrous. In the event, they closed within ¾ of an inch of each other.

In all, 11 TBMs were used, costing around £50 million in total. Shaped like huge cylinders several times a man's height, they weighed up to 1,500 tons and were as long as two football pitches. With revolving 'mouths' of tungsten teeth, they cut through the undersea chalk at a rate of up to 15 feet per hour, and the spoil they excavated from as far as 246ft below the seabed was used to reclaim 90 acres of land from the sea on the English side alone.

These huge moles were followed by teams of workers installing pre-cast, semi-circular, 5ft-thick concrete sections to create the walls of the tunnels. Over 420,000 of these segments were installed by the British teams, and more than 250,000 by the French. It was hot, sweaty and dangerous work. The nightmare scenario, of a major breach allowing millions of gallons of seawater into the tunnel, never materialised, but 10 men were killed during construction, eight of them British.

When it was finished in 1993 it was 31 miles in length, with 24 of those miles being under the sea; it remains **the longest undersea rail tunnel in the world**.

There are actually three tunnels – two 25ft diameter main chambers, with a smaller, 16ft serviceway and many linking passages to allow for work access, emergency escape and to relieve the air pressure barrelling ahead of speeding trains. These create 120°F temperatures through air friction as they travel at up to 85mph, and around 300 miles of piping filled with 18 million gallons of cold water are installed to bleed off this heat.

More than 300 trains travel through the tunnel every day – 212 million passengers have now used it between Britain and France, and 50,000 tons of goods are transported through it daily. The recent opening of the high speed Channel Tunnel Rail Link means that it is now possible to travel from London's St Pancras Station to Paris in a little over two hours.

Despite these impressive numbers, the Channel Tunnel has never made a profit, and there have been other problems. Fires aboard trains have raised safety concerns, there are fears that it provides terrorists with a potential target and illegal immigrants have used it to walk to the UK. However, it is, beyond argument, a genuine masterpiece of civil engineering, recognised as such around the world.

CHOCOLATE

ULSTERMAN **Hans Sloane invented drinking chocolate**. He worked as a doctor in Jamaica where the national drink was a mixture of cocoa and water. It made him feel nauseous so he added milk and sugar to improve the taste. On his return to Britain in 1689, Sloane (1660-1753) began prescribing hot chocolate as a pick-me-up for his patients.

The world's first chocolate bar was made by an Englishman in 1847. For a long time, chocolate had been only a drink, but Bristol confectioner **Joseph Fry** (1826-1913) developed a new technique, taking cocoa powder, extracted cocoa butter and sugar and mixing them to produce a paste that could be moulded and hardened. His solid *Chocolat Delicieux a Manger* soon captured the imagination and within two years rival chocolate-makers Cadbury had a similar product on sale. Fry's Chocolate Cream is still on sale today, though his company merged with Cadbury in 1919.

CHOLERA

CHOLERA was killing thousands of people each year in Britain (and millions all over the world) in the mid-1800s. 'Bad air' was blamed for the illness which causes vomiting, diarrhoea and death if left untreated. But a Yorkshire labourer's son proved that it was a waterborne infection and, by cutting off the source, saved countless lives.

Epidemiologist **John Snow** (1813-1858) was a doctor in London in 1854 when cholera claimed 500 souls in Soho in just ten days. Snow suspected that contaminated water was the cause, and he ordered a map large enough to show every street, house and well in the district. By marking each death on the map, he proved that those who died lived near to the Broad Street water pump. He suggested the handle of the pump be removed, and afterwards new cases of the disease immediately diminished. (Another Briton, **William Budd**, had independently arrived at similar conclusions in Bristol, though he is best known for his work on preventing typhoid.)

A vegetarian member of the Temperance Society (though he started boozing and eating meat later in life when his health began to fail), Snow made a number of other important contributions to medicine. In 1841 he devised a pump that could be used for artificial respiration and in 1844 he came up with an instrument for removing fluid from the chest while avoiding the entry of air, so preventing the collapse of the lung. He also made anaesthetics safer by testing controlled doses on animals and humans, and was trusted enough to give Queen Victoria chloroform during childbirth.

CHRISTMAS

FATHER Christmas first appeared in 1616 in *Christmas his Masque*, a play by **Ben Jonson**. He was known simply as 'Christmas', however, and was portrayed as an elderly man with a hat and beard – he entirely failed to slide down any chimneys or leave presents under trees. The modern image of him in a red outfit dates from the Victorian era –

despite endless claims, Coca Cola did not invent him for their 1930 advertising campaign.

Christmas crackers were invented by Tom Smith in London in 1847, with the bang intended to imitate the crackling of a log fire. Coloured hats and awful jokes were added later. The **world's first commercial Christmas card** was made and sold in Britain in 1843. Henry Cole commissioned them and sold all 2,050 of his print run for a shilling each. The modern **Christmas Pudding**, with its distinctive cannon ball shape and iced top, first appeared in the 1830s. Elizabeth Acton mentioned it in her 1845 book *Modern Cookery for Private Families*, which was also one of the **first cookbooks** aimed at the domestic audience.

CLEAN AIR

IN 1956 we were the **first country to introduce a Clean Air Act** which brought in smokeless fuel and taller chimneys in response to industrial smog. It helped to improve the air in our cities and was later copied around the world.

CLOCKS

BEFORE the 17th century, there was no accurate way of telling the time. Sundials, hourglasses and rudimentary mechanical devices offered little more than a rough idea. Then, in around 1582, the Italian astronomer Galileo Galilei discovered that the oscillation of a pendulum – a swinging weight – could be used to mark the passing of time. The first pendulum clocks were made in Holland in 1656 but **the world's first accurate clock** was made in Britain. The 'longcase' or 'Grandfather' clock was pioneered by London's **William Clement** in 1670, using a device called an 'anchor escapement' which reduced the size of the pendulum's 'swing' and, in turn, ensured the clock kept better time. This basic principle remained the cornerstone of the art until the 1930s, when quartz timekeeping took over.

Pendulum clocks suffered from two main drawbacks.

The first was that variations in temperature caused the metal pendulum to expand or contract, altering the swing. Clocks would lose time in the summer, for instance, as the pendulum expanded and slowed down slightly. This was overcome by English clockmaker **George Graham** from Cumberland, who used a glass cylinder partly filled with mercury to compensate (as the cylinder lengthened, the mercury expanded in the opposite direction, keeping the pendulum balanced). This improved accuracy to a loss of only one second a day (as good as early digital watches in the early 1980s). **John Harrison** (see *Longitude Problem*) came up with an even better solution in 1761; he built a pendulum out of bars of different metals, whose different expansion rates cancelled each other out. This further increased accuracy fivefold.

The second problem was that they were based on the assumption that the pendulum swings completely freely, which is impossible in practice. Friction in the mechanism, air resistance and vibrations all introduce inaccuracies. In 1921, railway engineer **William Hamilton Shortt** overcame this by having two pendulums; one was kept in a vacuum-sealed container and sent an electrical signal to the other which was used to drive the clock. His Shortt clock lost less than a second per year. For two decades it was used in laboratories throughout the world and was unsurpassed in accuracy until the introduction of the electronic clock.

The first electric clock was built in 1840 by **Alexander Bain** (1811-1877, see *Fax Machine*), and **the world's first accurate atomic clock** was built by Nottingham-born **Louis Essen** in 1955 at the National Physical Laboratory in Teddington, London. Atomic clocks use the microwave signals emitted by the electrons within atoms to mark time: Essen's used caesium-133 and would lose only one second every 300 years. He was the only British physicist honoured for his contribution to science by both the USA and USSR during the Cold War.

By the way, Salisbury Cathedral Clock is the oldest working clock in the world, dating from 1386. It is made from iron, has no dial and strikes each hour.

CLOCKWORK RADIO

A FORMER underwater stuntman and international swimmer, **Trevor Baylis** began work on his wind-up radio in 1989. He spent 18 hour days working on the invention in his garden shed at home on Eel Pie Island, Twickenham. His aim was to bring mass communication to the poor in far-off lands, to make it easier to spread educational information about AIDS. He later said: 'The key to success is to risk thinking unconventional thoughts. Convention is the enemy of progress. As long as you've got slightly more perception than the average wrapped loaf, you could invent something.'

CLONING

DOLLY the sheep became a scientific wonder and an unlikely celebrity when she was revealed to the world in 1997.

Created at the Roslin Institute, near Edinburgh, she was the **first mammal in the world to be cloned** from a mature animal – this meant that her mother was also her identical twin.

It took a team of scientists led by **Dr Ian Wilmut** 277 attempts to create the healthy lamb. They took the nucleus out of the centre of a donated egg and inserted DNA from a single cell taken from a ewe's udder. It was kick-started, Frankenstein-style, with a tiny electrical impulse and once the cells started multiplying, the embryo was implanted in the surrogate mother.

Born on July 5, 1996, Dolly, a Finn Dorset white sheep, was genetically identical to the ewe who donated the cell and unrelated to the surrogate, a Scottish Blackface ewe. She was named after singer Dolly Parton. Dr Wilmut, from Hampton Lucy, Warwickshire, said, 'Dolly is derived from a mammary gland cell and we couldn't think of a more impressive pair of glands than Dolly Parton's.'

Dolly went on to have six healthy lambs of her own – produced the conventional way – before being put to sleep in 2003, aged six. She was suffering from arthritis and lung disease; the latter is common in older sheep, and it's possible that because the original cell was taken from a six-year-old sheep, Dolly's DNA was 'older' than she was. Following her death she was stuffed and put on display at the National Museum of Scotland, in Edinburgh. She was produced as a step towards creating animals that will be genetically modified to produce medicine in their milk, or create organs for transplant. Since her birth, cloned pigs, cows, cats, dogs and even a monkey have been created, and the first companies offering to clone dead pets have been set up.

CLOUDS

THE strange skies of 1783 changed the life of schoolboy **Luke Howard**, and gave us **names for cloud formations used around the world** to this day.

The Oxfordshire 11-year-old spent hours staring at the roiling clouds and bright red sun which filled the horizon for much of that year – caused by thousands of tons of ash and gas thrown into the atmosphere after the dramatic eruption of Laki, a volcano in Iceland.

Howard, who was born in London in 1772 and educated in Burford, trained as a chemist but never lost his keen interest in meteorology. Over 20 years, he whiled away leisure time sketching skyscapes and defining the various types of cloud. Until then, they were broadly viewed as transient 'essences', described by their colours alone. (A Frenchman, Jean Baptiste Lamarck, had tried to classify the main types of clouds shortly before Howard, but his classifications were in French rather than Latin, the language of science, so they were disregarded.) Howard came up with the following names:

- Stratus, meaning 'layer', for horizontal, flat, blanket cloud coverage

- Cumulus, meaning 'lump', for white, fluffy clouds

- Cirrus, meaning 'wispy', for feathery high clouds

- Nimbus, meaning 'rain', for clouds from which rain, hail or snow falls.

He also added sub-categories, such as 'cumulonimbus', for large lumpy clouds that produce rain, and 'altocumulus', for those which look fluffy or rolled but have dark, shadowed sides or a dappled appearance.

COMPUTERS

INVENTIVE British minds played a huge role in the development of computers.

The ancient Chinese abacus was the earliest 'computer', but for our purposes we start with **John Napier**. Napier was born in Edinburgh in 1550. He was a highly intelligent man regarded by some as a magician for what we would now call lateral thinking. For instance, when one of his servants stole a trinket from his house, he lined them all up and had them stroke his black cockerel. He had explained that it would crow when stroked by the thief. It didn't crow, of course – Merchiston's methods were more prosaic. He had smeared the rooster with soot, and the guilty man was identified by his clean hands – he'd been afraid to stroke the bird for fear of discovery. (Nowadays, the law unfortunately prohibits this type of questioning.) On another occasion, fed up with his neighbour's pigeons constantly eating the grain in his fields, Napier warned the man that he would henceforth keep every pigeon that ventured onto his land. The neighbour shrugged, knowing that it was virtually impossible to catch even a single pigeon. The next day he was astonished to see Napier walking around his fields casually scooping up pigeons into a sack; he had soaked peas in brandy overnight and fed them to the birds, who became too drunk to fly.

Napier's giant contribution to the birth of computing grew from his interest in astronomy. His star-gazing required complex and time-consuming calculations involving very large numbers. He devised a system – which he called '**Logarithms**' – that enabled the multiplication and division of large numbers to be reduced to simple addition and subtraction. The printed tables in his 1614 work

A Description of the Wonderful Canon of Logarithms speeded them up still further. Centuries later, computers would use logarithms – and 'log' tables themselves remained in widespread use for more than 350 years, until the invention of the electronic calculator. He died of gout in 1617, having also invented a calculating device made up of a set of rods which became known as Napier's Bones. This was the first major improvement to the abacus and enabled large numbers to be easily multiplied or divided without any additional tables or particular expertise. Additionally, he popularised the use of the decimal point.

William Oughtred (1575-1660), a Buckinghamshire-born vicar and mathematician, invented the **Slide Rule** in 1622. This revolutionary hand-held device used the mathematical properties of the logarithm to further speed up long multiplication and division. Unfortunately, it wasn't until the mid 19th century that engraving and machining caught up with his brilliant design so that devices of a high-enough quality could be fabricated to make the most of its potential. But once they did, the slide rule became the universal symbol of the engineer, as the stethoscope is for the doctor. In the hands of a skilled user, the slide rule was a very speedy means of carrying out complex calculations, and they were even carried as a back-up by astronauts on the Apollo moon missions. In the 1970s, they were widely replaced by electronic pocket calculators, but Oughtred's slide rule is still not quite dead; pilots still have to know how to navigate with 'dead reckoning', using a circular slide rule.

Of course, logarithms and slide rules were but early steps on a very long road.

Two centuries after Oughtred came **Charles Babbage** – as archetypal an eccentric genius as Britain has produced. Born into a well-to-do London family on Boxing Day in 1791, he survived near death from fever twice as a child and excelled at maths – like Sir Isaac Newton and Stephen Hawking, he was Lucasian Professor of Mathematics at Cambridge. An irascible and impatient man with a strange hatred of organ grinders and a wide-ranging mind, he invented the cow catcher, used on steam trains to scoop wandering cattle off the tracks, had himself baked in an oven for five minutes to see what would happen and in 1847 made **the world's first known ophthalmoscope**, the device used for examining the interior of the eye. (No-one was interested, and credit went instead to a German doctor, Hermann von Helmholtz.) He was obsessed with facts and accuracy: his biographer, Doron Swade, tells how Babbage (with tongue in cheek) sent a letter to Tennyson to complain about his poem *The Vision of Sin*. 'In your otherwise beautiful poem,' wrote Babbage, 'one verse reads, "*Every moment dies a man, Every moment one is born.*" If this were true, the population of the world would be at a standstill. In truth, the rate of birth is slightly in excess of that of death. I would suggest (that you rewrite the poem to read): "*Every moment dies a man, Every moment 1 1/16 is born.*" Strictly speaking, the actual figure is so long I cannot get it into a line, but I believe the figure 1 1/16 will be sufficiently accurate for poetry.'

That obsession with accuracy left him infuriated with the 'computers' of his day. In the early 1800s, a 'computer' was a clerk who sat in a gloomy, Dickensian booth, with a pencil, a book of tables and maybe a slide rule, and spent all day doing difficult, boring sums. He might be calculating interest repayments on loans, analysing the stresses on a wooden beam or determining a ship's position at sea: the key thing is that he was a human being using charts produced by other human beings. Boredom, tiredness and repetition all had the potential for error, both in the tables used for calculation and in the calculations themselves. If a clerk made an error in an interest rate calculation, the debtor might underpay (or overpay) but other errors might lead to artillery shells falling on friendly troops, bridges collapsing or ships running aground.

In 1821, while checking some newly-computed astronomical tables, Babbage found dozens of mistakes. He is said to have exclaimed, 'I wish to God these calculations had been executed by steam.' He decided to eliminate human error by eliminating the human. In 1823, he unveiled designs for a huge '**Difference Engine**' to an enthusiastic audience at the Royal Society, and won an initial government grant of £1,500 (some £4 million today, as a proportion of GDP) to start work on its manufacture.

Actually first conceived by the German engineer Johann Müller, the Difference Engine (or engines; Babbage refined his design on a number of occasions) was so-called because it was intended to calculate finite differences using addition (which was simpler to mechanise than multiplication and division). The idea was that it would take all the work hitherto done so painstakingly by hand and pencil and crank it out automatically though a system of tiny gears. As such, it was **the world's first mechanical computer** – though it was more akin to a giant, inflexible, fixed-function calculator, able to add up to many decimal places.

Babbage, his personal life stalked by tragedy, never saw it built. In 1827, his father, wife and two children all died and he took a year off to recover. That, together with funding difficulties and squabbles with his master engineer, led to its effective abandonment in 1832. By then he had spent £17,478 (£50 million) of the government's money. At a meeting to seek more state aid, Prime Minister Sir Robert Peel suggested sarcastically that the best use for the machine might be to calculate the date when it would be of some use to the country. Peel was not alone in simply being unable to envisage the potential it offered.

Undaunted, Babbage started work on designs for what he called his **Analytical Engine**, outlining these plans in 1837. If his Difference Engine was a dramatic advance, this 'second generation' device was genuinely breathtakingly in scale and ambition (though, again, its design evolved over time, so 'Engines' is a better description). Powered by steam, it would have had 25,000 parts, stood 8ft high and 90ft in length, and weighed around 15 tons. Anticipating modern computers, it was to be programmable (by feeding in punched cards) and would have a printer for issuing results. Its 'store' was capable of holding

1,000 numbers of 50 decimal digits each (the equivalent of more than 20kb of modern memory) and its 'mill' – effectively, the CPU – was to perform addition, subtraction, multiplication and division, together with other more complex problem-solving.

It's beyond the scope of this book* to delve much deeper into the intricacies of Babbage's Analytical Engine (readers anxious to know more about the maths behind and inside the Engine[s] can find it in works by Doron Swade, Betty Toole, Anthony Hyman and others), which he worked on until his death in 1871. Again, it was never finished – or even properly started, beyond many detailed drawings – and it's not hard to see why. The technical requirements involved in making the parts and constructing the whole exceeded the limits of 19th century engineering, and the finances required, which he intended to supply himself this time, were huge.

After his post mortem, his brain was pickled and then left in a jar on a shelf, gathering dust. It was finally dissected in 1908, by which time he was all-but forgotten. In 1985, however, the Science Museum decided to build his second great Difference Engine (from his 1847-9 designs). It took six years, using only the techniques that would have been available to Babbage, and has over 4,000 working parts. The finished product did what he had predicted it would, and is on show at the museum. While the development of all of his Engines was beset by failure and heartbreak, they represented true, visionary brilliance.

In 1936, the next great leap forward in computing came in a paper written by **Alan Turing** (1912-1954), a genius from Maida Vale who is considered the **father of modern computer science**. Again, its complexity is beyond the scope of this book (as is a fuller history of computing): suffice it to say that it showed that machines could

---

* The British contribution to computers and computing is much wider than we have space for. Other key figures include Michael Faraday, who discovered how to generate electricity using magnets, George Boole, the originator of switching theory, and Henry Mill who patented the typewriter, crucial to any modern PC, in 1714.

theoretically be built which would answer any mathematical problem (though the state of the art at the time meant that any contemporary machine would be extremely limited).

As so often happens, the outbreak of a war – in this case, WWII – had a dramatic ratcheting effect on progress. Within months, German submarines were sinking Allied ships with virtual impunity. We needed to discover their patrol locations to destroy them, but the enemy military used a supposedly uncrackable system called 'Enigma' for its communications. Turing's brilliance was turned towards cracking the complex German ciphers at the main British codebreaking centre at Bletchley Park, Bucks. Although he and his colleagues made some progress, the breakthrough came in March 1941 when a German ship, the *Krebs*, was captured off the Norwegian coast with two working Enigma machines on board. Other machines and code books were captured from enemy weather and supply ships and from April 1941 the Bletchley Park staff were able to decrypt Enigma transmissions.

Turing and his comrades also cracked 'Lorentz', a fiendishly difficult system used only for the most important communications from Hitler's High Command. Messages were typed in to a device which added a seemingly random stream of characters to render the output indecipherable to anyone but the recipient, who was using a similar machine. The code breakers took advantage of a single operator mistake made in 1941, where a message was sent twice using the same key positions. From this single intercept, the original message, and the stream of characters added by the machine, were identified. This error enabled the engineers to construct a copy of a Lorentz cipher machine. They had also identified a weakness with the Lorentz machine, showing that its output was not quite as random as the Germans believed it to be; with enough output examples, a pattern could be deduced and the original messages unscrambled. To analyse the messages, however, they needed a computer.

In 1943, the GPO engineer **Tommy Flowers** (1905-1998) designed the Colossus Mark 1 with help from the mathematicians at Bletchley. It was a room-sized construction of vacuum tubes built specifically to break the Lorentz cipher. Colossus was **the world's first digital, electronic and (partially) programmable computer**. A Mark 2 version five times faster followed in 1944. (The project was codenamed 'Ultra'; the Germans never discovered it, and many historians believe the work carried out by Turing, Flowers and others in the 9,000-strong staff shortened the war by at least a year.) Turing died in 1954, at the tragically young age of 41, after eating a cyanide-laced apple. Most people believe he committed suicide, though some think he may have been assassinated by the Security Services who feared his homosexuality might leave him open to blackmail and possible coercion into spying.

The Colossus machine remained top secret: after the war, much of the equipment was destroyed and the blueprints burnt on the orders of Churchill. Indeed, information about Colossus did not become public until relatively recently, so it does not feature in many of the early histories of computing. In 1994, a team started rebuilding a working Colossus Computer, and it was demonstrated in November 2007.

Britain maintained its early lead by producing **the world's first computer able to store its own program** at Manchester University in 1948. This led directly to the first commercially-produced machine, the Ferranti Mark One in 1951. Unfortunately, it was not a lead we maintained for long, despite the lightbulb moment enjoyed by British visionary **Geoffrey Dummer**. The Hull-born engineer worked on radar during and after the war, and wanted to improve the reliability of electrical components. In 1952, he made a presentation to the Electronic Components Symposium in Washington DC in which he outlined his idea of the possibility of etching the different parts of a circuit into a single piece of material. This 'Integrated Circuit' was **the first microchip**. Sadly, Dummer struggled to get any interest or investment from either private companies or the British military, and it was left to American Jack Kilby and his forward-thinking employers at Texas Instruments to build and patent their own integrated circuit in 1959.

In 1972, **Sir Clive Sinclair** produced **the world's first slimline pocket calculator**, the 'Executive', and endeared himself to students of mathematics the world over in the process. In 1980, he endeared himself to their younger brothers and sisters by launching the ZX80 personal computer. Although other countries and firms – notably the USA's Commodore – had already produced home PCs, Sinclair's genius was in pricing. The ZX80 cost £99.95 (cheaper still if self-assembled), a fifth of the cost of a Commodore. It sold in its tens of thousands, and was followed in 1981 by the improved ZX81 which sold more still. But it was Sinclair's ZX Spectrum, revealed in 1982, that really brought the world of computers into British living rooms. Retailing at £125, it was much cheaper than the opposition and, for the first time, the average family considered buying a computer – previously they had really only been of interest to electronics enthusiasts and businesses. The machine was expected to sell a thousand a month, but at one stage Sinclair found himself selling 15,000 a week as children up and down the land begged their parents to buy one 'to help them with their homework'. The Spectrum is also credited with kick-starting the computer games industry, and many of today's programmers grew up tinkering with programmes for it.

Sinclair made a fortune, was knighted and promptly launched the C5 electric car. This was a road-legal, battery-powered one-seater white tricycle which could reach 15 mph. The driver's head was at around the same height as the top of an HGV wheel. Unsurprisingly, it was a terrible flop.

CONCORDE

CONCORDE is **the world's only successful supersonic airliner**.

Engineers around the world had long dreamed of building a passenger jet which could fly faster than the speed of sound (around 700mph, depending on air temperature and density). Her arrival forced the Americans to admit defeat and drop their own plans for a supersonic aircraft. Although the Soviet Union did develop the uncannily similar-looking Tu-144, that was a far less successful plane and may well have been developed from stolen plans.

Concorde, a graceful, swept-wing machine derived from a 1961 Bristol Aeroplane Company design for a six-engine jet which would cruise at well over 1,000mph, was developed jointly with the French. Her first flight was a 20-minute subsonic hop from Bristol to RAF Fairford, Gloucestershire, in 1969. After landing, test pilot Brian Trubshaw, a gregarious Welshman, climbed out of the cockpit and declared: 'It was wizard – a cool, calm and collected operation.'

The maiden commercial supersonic flight, from London to Bahrain, took place on January 21, 1976. Powered by Rolls-Royce Olympus engines based on those built for the Vulcan bomber, Concorde flew at Mach 2.2 (1,450mph) at a ceiling of 60,000ft and cut the journey time between Heathrow and New York in half. Travelling west, time differences meant that she could land technically earlier than she had taken off. Travelling eastbound, passengers actually weighed around 1kg less during the flight due to the centrifugal force generated by the aeroplane's speed combined with the Earth's rotation.

Her supersonic speeds threw up many engineering problems. Chief among these was the heat caused by air friction: the cockpit windows became hot to the touch and the airframe stretched by up to a foot at Mach 2, causing a gap to open up on the flight deck behind the pilot's seat which closed up again as the plane slowed down; in order to radiate heat from the aluminium airframe, it had to be painted white.

The Concorde project cost six times its original budget, and was a financial white elephant. Initial interest and orders from airlines around the world collapsed after the price of oil rose in the early 1970s and the loss of a Tu-144 at the 1973 Paris Air Show in a crash which killed 14 people.

In the end, only 20 Concordes were ever built, at a 1980 cost of around £23 million each, but it was perhaps the most beautiful aircraft ever to fly, and it set records that will stand for decades to come. On 25th July, 2000, Air France Concorde Flight 4590 to New York crashed during takeoff from Paris, killing all 109 people aboard and four on the ground. The official report said the aircraft had hit debris on the runway, causing a fuel tank to puncture and the fuel to ignite. The planes were grounded for modifications and did briefly fly again, but they were all retired in 2003 by British Airways and Air France.

CONCRETE

IT may look ugly, but without the work of two British building pioneers, many of the world's most impressive civil engineering projects would have been impossible.

The first is **John Smeaton**, who was born in Leeds in 1724 and **invented modern concrete**. Smeaton – regarded as the 'Father of Civil Engineering' – used it in his construction of the third Eddystone Lighthouse, off the coast near Plymouth, between 1755 and 1759. The first lighthouse had blown down in a great storm in 1703, and the second burnt down in 1755, despite the best efforts of its keepers to save it. (One of them, 94-year-old Henry Hall, claimed to have swallowed molten lead falling from the top of the tower during the futile efforts to douse the flames. He died five days later, and an autopsy revealed half a pound of lead in his stomach.)

Smeaton was determined that his lighthouse would fare better. Although the Romans had used some concrete, they had nothing as strong or durable as Smeaton's modern version, which involved adding pebbles and powdered brick into a hydraulic lime cement. (Hydraulic lime is produced by heating limestone containing clay and other impurities to a temperature of over 1,000°C. This causes a chemical reaction between the calcium in the limestone and the impurities, creating the basis of cement powder.) Crucially, Smeaton's concrete could set underwater; his lighthouse, made with granite blocks dovetailed like woodworked joints, stood for 120 years until the rock *below* it started to erode, at which point it was dismantled and rebuilt at Plymouth Hoe.

Incidentally, it was he who coined the term 'Civil Engineer' to distinguish men like himself from the military engineers who were then normally responsible for public infrastructure works. Smeaton designed numerous bridges and canals, and died in 1792.

The second Briton is **Joseph Aspdin**. Also from Leeds, he was born in 1788 and his contribution was to **develop Portland cement** which he patented in 1824. Until then, lime was the main ingredient of mortars and concrete, but it was a difficult material to work with – it required slaking with water, a process which took up a great deal of time and space. Portland cement came ready for use – the builder simply mixed one part cement with three parts sand and added water and the resultant mortar was hardened within hours. The speed involved mean construction was vastly quicker.

The cement – so-called because it looks, on drying, similar to Portland stone – is made from stone containing oxides of calcium, silicon, aluminium, iron and magnesium. Hewn rock is smashed to pebble size, heated and then ground into powder, with further chemicals being added as required to produce cement with the required properties of strength, setting speed and so on. It is now used in 90% of building work around the world.

An honourable mention should go to **William Boutland Wilkinson**. Born in 1819 in Newcastle, he **pioneered reinforced concrete** – the process of inserting metal strengthening bars into poured concrete to provide additional strength – in 1854. Without this breakthrough, all the Portland cement in the world would be useless in the construction of skyscrapers, bridges and most other modern developments. Unfortunately, although he built the world's first reinforced concrete houses in the north east, he didn't bother to patent his process; he left that to the Frenchman Joseph Monier in 1867.

THE CONTRACEPTIVE PILL

A CRYING baby was perhaps the commonest form of birth control until 'the pill' was launched in the 1960s. A British bookmaker's son is behind **the world's first cost-effective oral contraceptive**.

Herchel Smith was born in Plymouth in 1925. He studied at Cambridge and Oxford, and then lectured in organic chemistry at Manchester University, where he devised and patented ways of creating new steroids. After moving to the USA, he discovered an inexpensive method of manufacturing 'norgestrel', an artificial hormone which prevents conception. This paved the way for practical birth control – earlier pills were hideously expensive because they used hormones extracted from natural sources. Ovral, the first entirely synthetic contraceptive, was based on his work and became available in 1968.

Smith held 800 patents for his work and they made him a multi-millionaire. A keen art collector who spent his retirement cruising around the Caribbean in a yacht called *Synthesis*, he was an astonishingly generous benefactor and philanthropist: on his death in 2001, he left £45 million to Cambridge University, its biggest ever legacy, and millions more to Harvard.

CONTROLLED CLINICAL TRIALS

NAVAL surgeon **James Lind** (1716-1794) carried out what is believed to have been **the world's first controlled clinical trial** in 1747. In the process, he proved that citrus fruit prevented scurvy.

Born in Edinburgh, the merchant's son started his medical career as an apprentice to a ship's surgeon. He saw first-hand the effects of scurvy on sailors – ulcers, bleeding, loss of teeth and hair, the opening of old wounds, tiredness, nausea and, eventually, death. Now known to be caused by a vitamin C deficiency, at the time it was a mystery, and was a huge problem for a nation whose maritime horizons were steadily expanding.

Lind maintained that scurvy caused more deaths in the Navy than enemy action, and George Anson's disastrous attempt at circumnavigating the world between 1740 and 1744 highlighted the problem – nearly 1,000 out of 1,400 men on that voyage died of the disease.

When Lind became ship's surgeon of *HMS Salisbury*, he carried out an experiment with 12 scurvied sailors. He divided them into pairs and gave each pair a different potential remedy – either cider, elixir of vitriol (also known as aromatic sulphuric acid), vinegar, sea water, oranges and lemons or a spicy paste made with garlic and mustard seed. Unsurprisingly, the pair who had citrus fruit recovered (until they ran out of fruit) and the others did not. Unfortunately, vitamins were not understood at the time. The significance of Lind's experiment was not recognised and the Navy did not take up his suggestion to supply sailors with fruit; it was 40 years before lemon juice was supplied to ships and scurvy almost eradicated from the Royal Navy.

Lind made a further contribution to sea travel when he discovered that condensed steam from sea water (ie distilled water) could be safely drunk. He proposed that ships used this to supplement their fresh water stores. This policy was not adopted until after his death as there was no practical method to boil sufficient quantities of seawater until the early 1800s.

CORD BLOOD TRANSPLANT

IN 1988 the life of a five-year-old boy with a rare genetic illness was saved by a new kind of transplant operation – thanks to the pioneering work of **Edward Boyse**, a British professor who recognised that a widely-available waste product, incinerated daily in hospitals all over the world, had life-giving properties.

Professor Boyse (1923-2007), an immunologist from Worthing in Sussex, realised that blood from the umbilical cord and placenta is rich in stem cells (see *Stem cells*) and could be used to treat a variety of ailments. He thought it would be most useful in treating cancer patients whose bone marrow and stems cells had been damaged by chemotherapy.

Following his research, **the world's first cord blood transplant** was given to a French boy who was suffering from Fanconi's anaemia – a potentially fatal disease which affects the blood and bone marrow. The donor was the boy's baby sister. He was cured and, since then, more than 10,000 children born with fatal inherited blood diseases have been saved by a transplant of placental blood stem cells. (Cord blood is usually used in treatment of children because of the small quantity available from each birth.)

Dr Boyse, the son of a professional church organist, served in the RAF after leaving school and then studied medicine and moved to America. Among his other discoveries was the importance of scent in selecting a partner. Experimenting with mice – Boyse was an animal lover, and his lab animals were said to live in luxury – he found that they could smell the difference between relatives and strangers. They were more likely to be attracted to a mouse whose odour was dissimilar to their own, showing that they were not related and therefore not from the same genetic family.

THE CLINICAL THERMOMETER

UNTIL a Yorkshire doctor invented the clinical thermometer in 1887, having your temperature taken was no fun. Thermometers were a foot long and the patient had to hold one in his mouth or armpit for up to 25 minutes. **Sir Thomas Allbutt** (1836-1925), a vicar's son from Dewsbury, created an instrument that was six inches long and could take a reading in five minutes. He wanted a device that 'could live habitually in my pocket and (be) as constantly with me as a stethoscope' and adapted and miniaturised the mercury-in-a-glass-tube method invented by German physicist Daniel Fahrenheit. The portable thermometer was quickly adopted by other doctors.

THE CORKSCREW

IN the 18th century, wine, medicines and cosmetics were routinely stored in containers sealed with corks, as improvements in glass manufacturing made bottles cheap and strong.

Of course, once you've stuck a cork in a bottle you have to find a way to extract it. **The first known patent for a corkscrew** was granted to English clergyman **Samuel Henshall** of Christchurch, Middlesex, on August 24, 1795. Between then and 1908, perhaps partly reflecting our ongoing national interest in alcoholic beverages, nearly 350 British patents were awarded for corkscrew design. Birmingham, with its skilled workers and expertise in engineering, dominated the field worldwide. According to the 1858 *Dix's Directory of Birmingham*, there were then 16 businesses in the city making corkscrews. In 1888, **James Heeley** was granted a patent for his double winged corkscrew – the one with levers either side which is still commonly used today – called the 'A1 Heeley Double Lever'.

CRICKET

THE Americans, in particular, look down on cricket: how can it be a genuine world game when it's really only played in England and a few former British colonies, and you stop for tea? Given that their alleged 'World Series' involves 29 US baseball clubs and just one team from another country (Canada), this is a bit cheeky, but it's also wrong. Although cricket originated in England (in the 1500s), it's played enthusiastically in Scotland, Ireland and Wales – and in most other countries around the world. The list of current associate and affiliate members of the International Cricket Council is a genuine A (Argentina) to Z (Zambia) of exotic and far-flung locations. In the last couple of years, you could have watched Afghanistan beat Jersey by two wickets in the final of the ICC World Cricket League Division 5 (Hasti Gul Abed steered the Afghans home after they collapsed to 42 for 7 in the face of some hostile bowling by the Channel Islanders), or China defeat Myanmar (Burma) by 118 runs in a game described on the 'Cricinfo' website as 'the battle of the dictatorships'. You could even have seen Russia's national team play its very first match – in Moscow, against a touring side put up by Carmel and District CC from North Wales (Carmel were thrashed, but then Russia's population is estimated at 140,702,096 while Carmel's is 'more like 500 or so').

Major Test-playing nations include Australia, South Africa, Pakistan and, above all, India – where approximately 100 million people play the game. The first rules were drawn up in 1774, and the first Test match was between England and Australia in 1877 (Australia won), and the greatest ever sportsman – in terms of his statistical dominance over all rivals – was a cricketer. Sir Donald Bradman, the Australian batsman, averaged 99.94 runs per Test innings – more than 39 runs ahead of his nearest rival, Graeme Pollock (Cricinfo, players who have completed at least 20 Test innings); this is like a sprinter finishing the Olympic 100 metres final three or four seconds ahead of the next quickest runner.

Pub quiz bonus: The first ever international cricket match was, slightly bizarrely, between America and Canada in 1844.

CROSSWORDS

WHAT did **Arthur Wynne** invent? Here's a clue: angry remark, nine letters.

Answer: The crossword.

Wynne (1871-1945) was born in Liverpool, the son of *Liverpool Courier* editor George Wynne. In the 1890s, he emigrated to America where he worked first as a Texas onion farmer before deciding to follow in his father's footsteps. Following jobs on newspapers in Ohio and Pennsylvania, he became editor of the 'fun section' of the *New York World*. On December 21, 1913, with a gap in the paper, he devised **the world's first crossword puzzle** – he actually called it the 'word-cross' – and was inundated with letters from readers who loved it. He never copyrighted the idea and, consequently, made no money out of it. But he did have the satisfaction of knowing that his moment of inspiration gave a great deal of pleasure – and no small measure of frustration – to millions of people.

THE CT SCANNER

BY August 1966, the Beatles had sold 150 million records worldwide. But when John, Paul, George and Ringo's screaming fans were shelling out for copies of *We Can Work It Out* and *Yellow Submarine*, they had – unknowingly – been funding development of **the world's first CT scanner**.

Back then, The Beatles' record label, EMI (Electrical and Musical Industries), dabbled in more than music – it was also a successful industrial research company. **Godfrey Hounsfield** (1919-2004), a brilliant engineer, invented the scanner while working for EMI – funded by sales of the Beatles' records.

The Nottinghamshire farmer's son joined EMI in 1951. Initially, he worked on radar and guided missiles, but he helped to build the pioneering all-transistor computer in the late 1950s and that whetted his appetite for computer-based science. However, the computer division wasn't profitable, so EMI so it sold it off in 1962 – the year the company (oddly, if you think about it) signed the up-and-coming Merseyside beat combo. Hounsfield transferred to EMI's Central Research Laboratory where he came up with the idea of the CT (computed tomography) scanner, also known as a CAT (computerised axial tomography) scanner.

A vital diagnostic tool, CT scanners are advanced X-ray machines which take multiple images of the target area using several beams sent simultaneously from different angles. The resulting picture is far more detailed than an ordinary X-ray and can be manipulated to give three-dimensional images of organs, allowing doctors to see the inside of the body, and soft tissues, without the need for invasive surgery. Originally developed to provide images of the brain, it's often used to detect tumours, abscesses and lung disease, as well as being able to look at internal injuries like a torn spleen or kidney.

Hounsfield's first machine was completed in 1968. His first trial involved scanning a pig's brain; the job took nine days to complete, and a 'high speed' computer took two and a half hours to process the 28,000 measurements taken by the scanner. EMI patented the invention and Hounsfield improved it with the help of development funds from the Medical Research Council.

The first human patient to be scanned with Hounsfield's prototype machine was a woman with a suspected brain tumour, in October 1971. The first EMI scanner was revealed to the wider world in 1972. It cost about £100,000 and by then the technology had advanced to the point where a scan took four minutes and the computerised image could be produced in seven minutes. By the end of the 1970s, the blurry images of the first scanners had been replaced with clear, high-resolution pictures.

In 1979, Hounsfield, who never married and lived for his work, received the Nobel Prize in Physiology or Medicine, which he shared with American physicist Allan Cormack who had independently described a similar technique. He was knighted in 1981 and, by the time of his death from heart failure in 2004, CT scanners were standard equipment in leading hospitals around the world. They are now being used in military hospitals in Iraq and Afghanistan where they have internet links allowing military doctors to consult specialists around the world about diagnosis and treatment.

DARWIN, CHARLES

WHERE did we *come* from? It's a question humans have asked since we became sentient, and the answer came from a great British visionary.

Charles Darwin's **Theory of Evolution** is the best explanation we have of our origins – it's now accepted worldwide, apart from in very primitive places such as Arkansas.

Darwin was born the fifth child of a wealthy doctor in Shrewsbury in 1809. He was sent to Edinburgh University in 1825 to train as a doctor, but he found it either dull (in lectures) or terrifying (he hated the sight of blood), and instead spent a lot of his time reading about natural history, and wondering about how life might have developed. He never graduated, and instead moved to Cambridge where he took a degree in Theology in 1831. Rather than enter Holy Orders, though, he then enrolled on a Geology course (a pattern emerges of a young man who is very keen to learn, but isn't sure what). On his return from a field trip one day, he found a letter proposing him for the post of unpaid 'Naturalist' and assistant to Captain Robert Fitzroy of *HMS Beagle*, who was shortly leaving on a two-year voyage to map the coastline of South America. Darwin's father, who would have to fund his son's passage, found Charles' seeming inability to settle to anything irritating, and it took a great deal of persuasion for him to allow the young man to join the *Beagle*. If he had not done so, our scientific understanding of nature might be very different.

In the event, the *Beagle* expedition lasted five years, as Fitzroy sailed on through the Pacific Ocean, under the southern coast of Australia and back to England via the Cape of Good Hope. Along the way, they called in at various ports and Darwin carefully noted geological features that showed the Earth had changed over the aeons – such as strata of seashells found far inland in the mountains of the Andes – and collected fossils and other specimens.

However, it was when the ship docked on the Galápagos Islands on September 16, 1835 that Darwin started to form his theory of the origins of species. Found in the eastern Pacific some 600 miles off the west coast of Ecuador, the Galápagos are a group of 20 or so islands which boast a spectacular and unique variety of flora and, particularly, fauna. Because the islands are so far from the mainland – and often one another – there are myriad species of creature which are to be found solely on one particular island, and nowhere else in the world. In effect, the Galápagos offered the young scientist a living laboratory untouched by human hand, in which he saw how various creatures had evolved to fit their surroundings. For instance, he noted that different types of iguana appeared to have developed from the same original species. Some had pointed tails and lived on land; others lived in water and had flattened tails which seemed almost to have been designed to help them swim. Other lizards on other islands had different characteristics again – it was as though each variety had been adapted to its particular landscape.

From this, Darwin derived his theory of Natural Selection, published in his 1859 book *On the Origin of Species by Means of Natural Selection, or the Preservation of Favoured Races in the Struggle for Life*. (Others had been thinking along similar lines for some time, notably the Welsh naturalist and explorer **Alfred Wallace**.)

Simply put, this says that all creatures are prone to mutation. Most mutations will weaken them, but some will confer an advantage which makes the creature superior to its predecessors and peers, and thus more successful. This successful mutation will be passed on in breeding (whereas unsuccessful ones will not). Take polar bears and brown bears; they share a common ancestry, and it's likely that polar bears were once brown but that chance mutations occasionally threw up white-furred cubs. In snowy conditions, this gave them an advantage – camouflage – that their brown-coated brothers did not have. They were better able to catch prey, and so the white fur gene was passed on and replicated in animals living in the Arctic circle, until this became the norm. By contrast, bears born white in North America were at a *dis*advantage – their prey could better see them coming in the woodland – and so they would not survive to breed. Eventually, two disparate species formed.

When *On the Origin of Species* was published, it caused controversy, particularly among religious leaders who saw it as an attack on creation by God, but today Darwin's ideas are generally accepted. Indeed natural selection has been witnessed as it happens. A famous example is that of 'industrial melanism'. Prior to the Industrial Revolution, most peppered moths were speckled white, with just a few being dark black. Where the smog of industry polluted the air and darkened tree trunks, the percentage of the peppered moth population which was dark increased to greater than 90%. In unpolluted areas, the peppered white form remained predominant. What was happening? It was simply that where the background upon which they sat had darkened, light-coloured moths were seen easily by birds and eaten. Where the background remained light, the reverse was true. The species was evolving through the predation of those unsuited to a change in conditions.

In 1871, Darwin published his 'follow-up' book, *The Descent of Man, and Selection in Relation to Sex*. In this, he made clear his belief that mankind was itself descended from a common ancestor with the great apes – something like the average *Celebrity Big Brother* contestant, perhaps. Going back further still, that *Big Brother* contestant, and every other animal on the planet, evolved from an original common ancestor – likely to have been a primitive single-celled creature which, over billions of years, became Us.

DIABETES

THE World Health Authority estimates that more than 300 million people worldwide will be diagnosed with diabetes by 2025. Thanks in part to the pioneering work of a Scottish biochemist, the majority of patients will lead a relatively normal life.

John James Rickard Macleod (1876-1935), a church minister's son from New Clunie, Perthshire, **discovered insulin** and a way of using it to treat patients who were dying from diabetes.

Insulin is a hormone produced in the pancreas which helps the cells convert glucose into energy. People with diabetes either do not make enough insulin or cannot use insulin as well as they should, allowing glucose to build up in the bloodstream. This can lead to serious health problems, including blindness, kidney failure, heart disease and death.

Macleod, who studied medicine at university in Aberdeen, believed that the pancreas was important in lowering blood sugar levels in the body. But it was not until 1921, while working in Toronto with Canadian scientist Frederick Banting, that he discovered that this organ secreted insulin and that the hormone could be given to diabetic patients as a therapy. Macleod and Banting proved this by extracting insulin-producing cells from dogs' pancreases – making them diabetic – and then re-injecting them with their own insulin, showing that it controlled glucose levels. It was a life-saving discovery that earned both men a Nobel Prize in 1923.

Macleod and Banting also worked on methods of removing and purifying the hormone. In 1922, they injected insulin into a 14-year-old diabetic, Leonard Thompson, who was near death. Their treatment prolonged his life by 13 years. In the same year they patented their method of extracting insulin from animals for the treatment of humans.

The most significant breakthrough in diabetes care in the years following Macleod's discovery came in the 1980s, when **the first biosynthetic human insulin** was created in Liverpool.

Scientists at Dista Products, a British subsidiary of the American pharmaceutical giant Eli Lilly, genetically altered *E. coli* bacteria in such a way as to make it produce insulin identical to that which develops naturally in the human pancreas. Called Humulin, and first prescribed in 1982, it was **the first human drug in the world to be created using recombinant DNA technology**. (Recombinant DNA is DNA which does not occur naturally but is created in the lab.) It's estimated that 95% of the insulin now used is made this way.

THE DICTIONARY

WHAT'S the difference between a 'fopdoodle' and a 'dandiprat'? If you have around £15,000 to spare, you can find out by purchasing an original copy of **the world's first proper dictionary** – published in England in 1755, and compiled by **Dr Samuel Johnson**. Alternatively, David Crystal's rather cheaper abridged reader is highly recommended.

There were dictionaries published before Johnson's, but his was the first to concentrate on defining everyday words for the lay reader – and the contribution it made to the understanding of language, and to communication, was immense.

The project took a decade to complete, contained 42,733 entries and ran to 2,300 pages. He worked alone, which led to the odd blunder such as his definition of a 'pastern' as 'The knee of an horse (*sic*).' In fact, it describes the area between the fetlock and the hoof. A woman asked him how he had come to make the mistake. Johnson's modesty and dry humour is shown in his reply: 'Ignorance, Madam, pure ignorance.'

It contains some fabulous definitions (a fopdoodle is a fool and a dandiprat an urchin) and among the most famous is that for 'Oats': 'In England, fed to horses. In Scotland, fed to the people.'

Lichfield-born Johnson (1709-1784) was a truly great Briton – an eccentric, a philosopher and a brilliantly perceptive writer and critic, who is said to have told a man who sent him a manuscript for his consideration that it was 'both good and original. But the part that is good is not original, and the part that is original is not good.'

THE DIESEL ENGINE

IN 1890, British mechanical engineer **Herbert Akroyd Stuart** took out patents for his compression-ignition engine. These detailed all the features of **the modern diesel engine**, but unfortunately all the credit went to the German inventor Rudolf Diesel, who patented his version two years later and gave his name to it in the process. A diesel engine uses the force generated by sudden compression to ignite its fuel, as opposed to an internal combustion engine which uses a spark. Akroyd Stuart (1864-1927), from Halifax in Yorkshire, also invented the first practical fuel injection system.

DIETING

DR GEORGE Cheyne was **the world's first diet guru** – despite himself being as fat as a hippo.

A farmer's son from Aberdeen, he set up a successful medical practice in Bath in 1718 and began treating the aristocratic and affluent who suffered from 'diseases of over-indulgence' and frequented the spa town to take the waters.

Cheyne (1671-1743) was highly critical of patients who failed to take responsibility for their own health, claiming that there was nothing more ridiculous than people 'perpetually complaining and yet perpetually cramming'. In 1724, he wrote *An Essay of Health and Long Life* – the *I Can Make Thou Thin* of its day – which extolled the virtues of exercise and fresh air, and warned against foreign and luxury foods. It also advised purging by making yourself sick.

Sadly, he didn't take his own advice: by that point, he weighed in at a colossal 32 stones and travelled in a specially-designed carriage, the whole side of which opened to allow him to clamber in and out.

Despite this apparent contradiction, his book flew off the shelves – it was eventually reprinted seven times, and made him one of the best-known physicians in Britain. But his own life at the time must have been hellish: ulcers blistered his legs, he had gout, he suffered headaches, depression and lethargy and was so short of breath that a servant walked behind him carrying a stool so that the eminent doctor could take a rest if he had to travel more than a few yards. Eventually, a milk and vegetable diet helped Cheyne lose two thirds of his body mass, recover his exuberance and make him, in his own words, 'lank, fleet and nimble'. Although the weight went back on over time, he would return to his milk and vegetable regime to lose the pounds. He died in Bath aged 72 'in full possession of his faculties to the last'.

DIVING

UNTIL the early 1800s, the only way to spend any length of time underwater was to use a diving bell – a bell-shaped, open-bottomed affair which was lowered into the depths from a boat. These had been used since the days of the ancient Greeks, mostly for collecting sponges from shallow sea beds. In 1685, Sir William Phipps employed one to salvage treasure worth millions from the wreck of a Spanish galleon in the West Indies, and five years later the English astronomer Edmond Halley (who discovered Halley's Comet) built a bell and sat in it, 60ft under the Thames, for over four hours – just to prove he could, really. Halley's design called for weighted barrels of air to be sent down from the surface to replenish that inside the bell, and although this system was improved with the development of **the first air pump** in the 1770s by the civil engineer **John Smeaton** (*see Concrete*), diving bells had only limited application – not least because you couldn't leave one while submerged for longer than you could hold your breath.

A man from Deptford called **Charles Deane** changed all of that, and in doing so pioneered diving. Deane (1796-1848) was a caulker (caulking is making timber-hulled boats watertight by filling the gaps between planks with pitch or other waterproof substances) at Barnard's Shipyard in London. He was interested in the problem of fighting fires inside a ship's hold and in 1823 he patented a 'Smoke Helmet' for use in smoke-filled areas. It was made of copper and was fitted into a collar connected to a cloak strapped tight around the body to keep fumes out. A bellows pumped in fresh air through a leather hose.

In 1828, he and his brother John converted the contraption into a diving helmet and suit, and by 1836 they had produced **the world's first diving manual**, *The Method Of Using Deane's Patent Diving Apparatus*. In the years that followed, the German-born but Soho-based engineer Augustus Siebe used the Deane designs to create divers helmets which would stay in use for another 100 years.

Why is this important? Firstly, the ability to work underwater allowed for easier salvage. In 1834, Charles Deane was able to recover 28 of the 100 cannon lost in the sinking of *HMS Royal George* at Spithead harbour. Although the ship lay in only 65ft of water, the cannon – which were very valuable – would otherwise have lain on the seabed. Secondly, it improved Britain's ability in civil engineering where projects to build in or over rivers and the sea were concerned.

The first SCUBA diving equipment (Self Contained Underwater Breathing Apparatus) was made in London in 1878 by **Henry Fleuss** (1851-1933) of British firm Siebe Gorman. This was a rebreather – a tank of oxygen with a rope soaked in caustic potash to absorb a diver's exhaled carbon dioxide.

The diver's nightmare – decompression sickness, or 'the bends' – was identified by Scots physiologist **John Scott Haldane** (1860-1936). This occurs when gas bubbles in the bloodstream, compressed by diving at depth, expand or 'decompress' as the diver surfaces. It can be fatal and is always serious. Haldane's groundbreaking 1908 paper (written with Arthur Boycott and Guybon Damant), 'The Prevention of Compressed-Air Illness', led to the development of staged decompression – this involves divers ascending slowly, and waiting at various depths to allow the gas to assimilate with the body. (It was based on Irish scientist Robert Boyle's 1667 work on compression in gases.)

DNA

THOUGH Charles Darwin had explained where we came from, no-one really understood our make-up. Three brilliant British scientists (and one New Zealander) came up with the answer when they produced **the first accurate model of DNA**.

DNA – Deoxyribonucleic acid – contains the genetic instructions for the development of all living organisms. In lay terms, it is like the body's blueprints – it contains the basic instructions needed to construct a living creature. In 1953, **Francis Crick**, from Weston Favell, Northamptonshire, **James Watson**, who was born to Scottish parents in America, and is the last surviving pioneer of the original team, and **Maurice Wilkins** (New Zealand-born but working in the UK) shared the 1962 Nobel Prize in Physiology or Medicine for their work in the field, most famously their discovery that the structure of

DNA was a helix. The fourth scientist, **Rosalind Franklin**, from London, had died of cancer four years earlier, aged just 37, and the Nobel rules do not allow posthumous prizewinners. This was particularly poignant, given that Franklin had struggled for recognition from her student days. She had graduated from Cambridge in 1941, but was only awarded a degree titular (a degree in name only) as women were not allowed to receive full degrees at the time. This changed seven years later, and in 1998 the University held a special graduation ceremony for the several hundred women still alive who had gained degrees there before 1948.

DNA FINGERPRINTING

BRITISH police solved a baffling double murder case with **the world's first use of DNA 'fingerprinting'**, using a revolutionary technique developed by a scientist at Leicester University.

In 1983, a teenaged girl called Lynda Mann was found raped and strangled on a deserted footpath near the Leicestershire village of Narborough. Three years later, a second 15-year-old, Dawn Ashworth, was murdered in chillingly similar circumstances in the same area. In both cases, semen samples were obtained from the bodies, although it was only possible at the time to use these for identifying blood group. They were tested and each sample came from the same group; given the other similarities in the cases, it was clear that the same killer had claimed both victims.

And the police soon thought they had their man – a local 17-year-old youth called Richard Buckland admitted the second crime under questioning, although he denied the first. In days gone by, Buckland would quite likely have been prosecuted and sentenced, perhaps to death, on the basis of that confession.

But Leicestershire Police had a new weapon in their armoury – DNA testing. After seven painstaking years of research at nearby Leicester University, geneticist **Alec Jeffreys** had recently developed a technique for distinguishing one person's DNA from another's.

Jeffreys, born in 1950 in Oxford, was (and is) an interesting character. He is said to be 'quite possibly the top Twister player on the staff at Leicester University, following his win at the 2006 Departmental Christmas party' and an appearance on *Desert Island Discs* revealed that he had been a Vespa-riding, parka-wearing Mod during the 1960s, before turning into a hippie for a while, and then finally buying a 350cc motorcycle and becoming a rocker. He'd also been fascinated by science ever since his father gave him a chemistry set when he was eight. His enthusiasm for setting off small explosions was not diminished when he was scarred on his chin by a splash of concentrated sulphuric acid, and he would spend weekends dissecting roadkill under his Victorian brass microscope.

Jeffreys – now Sir Alec, and a much-respected Fellow of the Royal Society – had his 'eureka moment' at 9:05am on Monday, September 10, 1984, when he located similarities and differences in the DNA of a technician and his family members. He immediately realised its enormous potential for forensic work and was soon co-operating with local detectives on the murders of Lynda Mann and Dawn Ashworth.

The first tests he carried out showed that the semen samples were, indeed, from the same man. The second showed that they were *not* from prime suspect Richard Buckland, who was released without charge. So who *had* killed the girls? In a further groundbreaking twist, 5,000 local men were asked by the police to give DNA samples, either blood or saliva. This programme took six months, and, disappointingly, no matches were found. But later a man was heard bragging in a pub that he had been given £200 to donate a sample while masquerading as his friend, Colin Pitchfork, a local baker. In September 1987, officers arrested Pitchfork at his home in neighbouring Littlethorpe, and his DNA was taken: it matched that of the killer and he admitted both murders at his trial the following year. He was sentenced to life imprisonment and will not be released until September 2017 at the earliest.

Since then, thousands of criminals have been convicted worldwide on the basis of DNA evidence – Alec Jeffreys' discovery is one of the key tools in modern crime fighting.

DROPSY

DROPSY sounds like a quaint and fairly harmless disease; in fact, it's a sign of advanced heart failure which causes the body to swell up with liquid and leads to an horrific death. Thanks to a British doctor's brilliant and painstaking work, however, it is not something we fear today.

Shropshire-born **William Withering** (1741-1799) came from a medical family, and learned about botany and medicine at Edinburgh University. In 1775, whilst working in Birmingham, he found that a patient who was given an extract from the foxglove plant by a relative seemed to gain some relief from the symptoms of dropsy. Withering spent nine years experimenting with extracts from different parts of the plant (not easy, when most of it is poisonous), testing their effects on 160 patients and documenting the plentiful and unpleasant side effects. He established that dried, powdered leaf worked best – and it shows how far ahead of his time he was that it took another 140 years before the active chemical constituents were identified. Medicines derived from members of the foxglove family are still used to treat some heart conditions.

THE DYSON

MANY of our great inventions and discoveries happened in the last century, and the one before that. **James Dyson** is proof that British creativity and ingenuity is alive and kicking today, too.

At face value, **the first bagless vacuum cleaner** may not appear to stand comparison with the development of the computer, the discovery of a cure for cholera or the unravelling of DNA – and perhaps it doesn't. But inventions are not all enormous in scale – some just make life easier. The 'Dyson' did that, and it also generated billions of pounds in export revenue for this country.

Sir James, who was born in Cromer, Norfolk, on May 2, 1947, had already invented the ballbarrow and the Sea Truck, a flat-bottomed, high-speed motor boat, when he hit on the idea of creating a 'Hoover' with no bag. His design uses 'cyclonic separation' to separate the dust from the air sucked into its cylindrical main chamber. Essentially, this involves spinning the air at such high speeds that heavier-than-air particles are thrown out, like a football placed on a turning roundabout, and collected for disposal.

Partly supported by his wife's salary as an art teacher, he spent five years making 5,127 prototypes, only to be rejected by all of the major manufacturers. So he set up his own company to make the Dyson, and is now said to be worth £1.1 billion. Among other inventions is the Airblade, a hand-drier which 'scrubs' dirty air from a public convenience and then blows it back out at 400mph, 'scraping' water from the user's hands. Dyson's own motto? 'Enjoy failure and learn from it. You can never learn from success.'

The British actually invented the original vacuum cleaner, too. More on that later.

EARTH, AGE OF

HOW old is the Earth? The question perplexed people for millennia but, by the start of the 20th century, biologists believed it must be at least several hundred million years old for evolution to have had sufficient time to work. Geologists measuring rock strata had come to similar conclusions, while the physicist Lord Kelvin* had calculated that the Earth must be between 24 and 400 million years old by assuming that it had cooled to its present surface temperature from a molten ball of rock. Other scientists, though, thought the planet must be much *younger* – the Sun would surely have burned itself out

---

* Kelvin – born William Thomson in Belfast in 1824 – was a genius of physics, thermodynamics and engineering and the youngest Briton ever to attend university (he was 10 at the time). His many achievements are covered in detail at the University of Glasgow's Kelvin Society website.

within a few million years? A means of making direct measurements was needed.

Ever since the discovery of radioactivity by the French chemist Becquerel in 1896, scientists had been trying to use the rate of decay of radioactive elements as a 'clock'. The idea was simple enough – find a rock sample which contains one of these elements and see how much of it has decayed, so that the age of the rock can be calculated. In practice it was far less easy. How do you calculate the rate of decay – or decide how much of a sample has actually decayed, or measure the trace amounts of material involved? **Arthur Holmes**, born in Gateshead into a modest background, excelled at school and won a scholarship at the age of 17 to study physics at Imperial College. A keen geologist, he ended up devoting his entire working life to calculating the age of the Earth. His first breakthrough, though, came while he was still a student. Holmes (1890-1965) refined a process developed by American scientist Bertram Boltwood which measured the decay of uranium into lead. During the final year of his degree in 1910, he tested rock from Norway and estimated it was 370 million years old. His paper was published by the Royal Society a year later, and was **the first solid proof that the Earth was at least several hundred million years old**.

Holmes then took a job in Mozambique prospecting for minerals, but after almost dying from malaria he returned to work at his old university. By 1920, he had a wife and son and, needing to earn more money, he accepted a job with an oil company in Burma. This was another disaster – his son died of dysentery and the company went bankrupt. He returned penniless, but managed eventually to get the post of Head of Geology at Durham University. (The grand title concealed the fact that he was actually the only person in the Department.) Many scientists could still not accept the possibility that the Earth could be so old and Holmes endured a great deal of criticism. However he had the conviction of his ideas and continued to refine his methods. By 1946 he had proved that the Earth must be at least three billion years old (modern techniques have now advanced this to around 4.5 billion years). Holmes was also a vocal proponent of the Theory of Continental Drift – the idea that land masses were moving on the earth's surface – and this only became accepted just before his death.

ECCENTRICS

BRITAIN'S had more than its share of eccentrics. Many of them were blue-bloods, and they reached their peak in the 19th century, probably because aristocrats in those days had the time and money to be eccentric.

In a very crowded field, **John 'Mad Jack' Mytton** perhaps takes the biscuit.

Born in 1796, he was a thrill-seeking gambler and all-round lunatic who downed six bottles of port a day and rode a bear into his dining room to impress his guests. He was sent to Westminster School with an annual allowance of £400 (£25,000 today). He spent more than double that amount in his first year, and was expelled shortly thereafter for 'fighting his teachers'. He was admitted to Harrow, but didn't last long there, either. He was home-tutored for a while and, after receiving a promise from his tutor that he would not have to read a single book, agreed to go to Cambridge. He sent ahead three pipes of port – equivalent to more than 2,000 bottles – but never graduated and probably never even attended.

Mytton inherited Halston Hall in Shropshire aged 21. He kept 2,000 dogs, 60 cats and stables full of horses, including his favourite, Baronet, who would lounge in front of the fire with him. He often went hunting with ribs broken from riding accidents, and was extremely fond of filbert nuts – *extremely*. He had them delivered by the cartload, and he and a pal once devoured 18lbs of them on a memorable carriage trip from London to Shropshire. The two men were 'knee-high with shells' when they arrived.

He became MP for Shrewsbury, but lasted half an hour in the House, finding it boring; he preferred to spend his time bare-knuckle fighting with miners, force-feeding his horses port and trying to jump hedges and gates in his carriage. One on occasion – learning that a passenger had never had a carriage accident – he exclaimed: 'Well, you must have been a dashed slow fellow all your life!' With that, he whipped his horses up a bank and overturned the gig on purpose. Luckily, the passenger survived unscathed.

Twice married – one wife died and the other left him – his disregard for money was legendary. He once lost thousands of pounds when it fluttered from his open carriage window after he fell asleep on the way back from Doncaster races. He borrowed heavily, often just to give it away to servants or 'friends' who were never seen again. One George Underhill lent him 'a large sum' and eventually, after sending Mytton many letters asking for repayment, arrived at Halston demanding repayment. Mytton's biographer and friend, Charles Apperley, records how Mad Jack charmed Underhill into accepting a letter to hand to a Shrewsbury banker who would, he assured him, settle the debt on his behalf. In fact, the banker also ran an asylum; the letter read: 'Sir: Admit the bearer, George Underhill, into the lunatic asylum. Yours obediently, John Mytton.'

Within 15 years he was broke, having frittered away the equivalent of millions of pounds, and was forced to flee to France to escape his creditors. On the way, he espied a pretty girl on Westminster Bridge and offered her £500 a year (£400,000 today) to accompany him. She stayed with him for two years, which seems almost worthy of a medal.

In France one night he developed hiccups. Apperley watched as he shouted, 'Damn this hiccup! But I'll frighten it away!' Grabbing a candle, he set fire to his nightshirt. As a servant beat out the flames, Mytton declared, 'The hiccup is gone, by God!' and reeled into bed. The next morning Apperley found him 'not only shirtless, but sheetless, with the skin of his breast, shoulders and knees of the same colour as a newly-singed bacon hog'. Mytton insisted on attending lunch that day, proudly repeating 'Can't I bear pain well?' in between fainting fits.

Eventually, he was forced to return to England through ill health, where he was promptly arrested and sent to debtor's prison. According to Apperley, he was a 'round-shouldered, tottering old-young man, bloated by drink, worn out by too much foolishness, too much wretchedness and too much brandy'. He died penniless in jail, aged 38.

ELASTIC

THOMAS Hancock created elastic in the early 1800s. Rubber is naturally stretchy, but Hancock, born in Marlborough, Wilts, in 1786, created a machine which combined scraps of rubber in a way that allowed it to be re-rolled into sheets with increased elasticity. Paranoid about the possibility of his invention being stolen, he referred to it mysteriously as his 'Pickling machine' and insisted on the utmost secrecy from those he worked with.

ELECTRICITY

IT'S easy to think of electricity as belonging only to the modern age, but the Briton who coined the word – or at least, its original form, 'electricitas' – was born in Essex in 1544 and carried out experiments with static electricity during the time of Henry VIII. **William Gilbert** published his findings (which were not all that advanced) in his book *De Magnete*. (What *was* advanced was his suggestion that the core of the earth was iron and that the poles were magnetic).

In the 1720s, a dyer and self-taught scientist called **Stephen Gray** discovered that electricity could be conducted hundreds of feet by some materials, such as hemp string or iron wire, but stopped in its tracks by others, such as silk. (The French scientist John Desaguliers, who made his name in London at the same time, named these 'conductors' and 'insulators'.) Gray – born in Canterbury in 1666 – also found that his wire lost its charge if it was allowed to touch the ground (ie it was 'earthed') and invented the marvellous 'Flying Boy' experiment, in which a small child was suspended in the air, given a good electric charge and attracted scraps of paper to his hands. This demonstrated the ability of the human body to conduct electricity; regrettably, Health and Safety legislation does not allow this experiment to be performed in schools anymore. Gray died in 1736, dictating notes about an experiment to his doctor whilst on his death bed.

Another major early figure in electricity was **Michael Faraday**. Although he received little formal education, he became one of the greatest experimental scientists in history, making huge advances in electromagnetism and chemistry. Born in Newington Butts, Surrey, in 1791, he started off as a bookbinder's apprentice and taught himself science after work. In 1821, he demonstrated the first simple electric motor. He established the concept of the magnetic field and showed that every time it moved, an electrical current appeared. This demonstrated the link between these two poorly understood phenomena and led him on to make **the first electric dynamo**, the Faraday Disc, in 1831. He was also the first to realise that the supposedly different forms of electricity (static, current flowing in a wire or charge stored in a battery) were all variants of the same thing. Faraday became such a renowned enthusiast and expert in this new field that William Gladstone, Minister for Finance, asked him what use electricity could be. His reply, which rings a bell down the ages, was, 'One day, sir, you might tax it.'

He also invented an early form of the Bunsen burner, discovered benzene (a organic chemical solvent which has been used, over the years, in the decaffeination of coffee, as a petrol additive and as an after-shave, until its carcinogenic properties became apparent) and the laws of electrolysis. Faraday even delved into nanoscience when he reported that miniscule particles of gold had different properties to larger ones. (The differing chemistry of nanometre-scale – 10^{-9} – objects is a very hot topic in modern science, as they are starting to be used in every product from cosmetics to tennis rackets.) The unit of capacitance (how much electrical charge a substance can hold), the *Farad*, is named after him, and he was an early environmentalist, writing reports for the government on the air quality in cities and the pollution in the Thames. He died in 1867 at the age of 76.

ELECTRIC CARS

THEY'RE supposedly the transport of the future – but **the first electric car** was made by a British inventor in 1884.

At least, recently-unearthed photos seem to show the 19th century inventor **Thomas Parker** sitting in an electric vehicle built in that year. The pictures were revealed this year by Parker's great-grandson Graham Parker, a 76-year-old former BBC weatherman. Parker was a renowned electrical engineer of the Victorian era – he electrified the London Underground, and also created overhead tramways in Liverpool and Birmingham (as well and coming up with the smokeless fuel 'Coalite'). He didn't press ahead with his electric car because of the arrival of the internal combustion engine – a far easier and more powerful vehicle technology – and died in December 1915.

ELECTRIC LIGHT

AS ANY American will tell you, Thomas Edison invented the light bulb in 1879.

The trouble is, Americans are so often wrong – it was actually invented by the British.

The story starts in 1802, nearly 80 years before Edison, when Cornish chemist **Sir Humphrey Davy** (*see Miner's Safety Lamp* and *Anaesthetic*) passed an electric current through a strip of platinum. The metal glowed hot and bright, though the illumination didn't last long. In 1809, Davy, who was assisted by Faraday, invented the arc lamp – **the first usable light** – after noticing that electric current which leapt across a gap in a circuit formed a bright light. The first such lamps were powered by high-voltage batteries, with a continuous fountain of sparks passing between two charcoal rods connected to the contacts. Dazzlingly powerful, they lit public areas including railway stations, Alexandra Palace, London's law courts and The Embankment in London.

James Bowman Lindsay, who was born near Arbroath in Scotland a year after Davy, demonstrated **the first *constant* electric light** in July 1835, at Thistle Hall in Dundee. It's not clear how it worked, but it was bright enough to enable him to 'read a book at a distance of one and a half foot'. According to the Royal Society of Edinburgh, 'He was the first person to discover how to create light by putting electricity through a vacuum.' (The significance of a vacuum was that the absence of oxygen prevented the filament from burning up.) 'Despite being in a position to revolutionise the world with the light bulb,' the Society adds, 'he failed to patent the idea.' Instead, Lindsay decided to spend the next 34 years of his life compiling a dictionary in 50 languages, which was promptly ignored. This is known as a *Sliding Doors* moment.

Warren de la Rue, born in Guernsey in 1815, added a filament – made of coiled platinum – to a vacuum and passed an electric current through it in about 1840. His bulb worked, but platinum was hideously expensive and this made his invention impossible for widespread use. (De la Rue also invented the first envelope-making machine in 1851.)

The first patent for a light bulb was granted to Englishman **Frederick de Moleyns** in 1841. Little is known about him (and it sounds as though he may have had some foreign ancestry), but the lamp used powdered charcoal heated between two platinum wires.

Sir Joseph Swan, a physicist who was born in Sunderland in 1828, knew that the key was a filament as durable as platinum, but cheaper. After much experimenting, he used carbonised paper and gave a demonstration of a working light bulb in 1860. The filament soon burned out but with a better vacuum pump to remove more oxygen, he received a patent for his incandescent lamp in 1878. By

1880, he had come up with a commercially-viable, longer-lasting lamp which would burn for 13.5 hours. In the same year, his house in Gateshead became the first in the world to be lit by electric light.

Across the Atlantic, Edison received a patent for his light bulb in 1879 – a year after Swan. It was of a very similar design, but while Swan was still making improvements to his lamp, Edison's bulb burned for 40 hours. (Soon afterwards, he had a bulb that would burn for 1,000 hours). Following a legal battle about patent infringement, the two great inventors merged companies in 1882 becoming Edison and Swan United Electric Company, which later became Ediswan. So, rather than being 'the inventor' of the light bulb, Edison, like Swan (and pretty much everyone else in history), improved on the work of other great thinkers, and came up with a better product.

Incidentally, in February 1879 Mosley Street in Newcastle-upon-Tyne became **the first public road in the world to be lit with an electric light bulb** – Swan's, not Edison's.

ELECTROMAGNETISM

THE British electrician **William Sturgeon** (1783-1850) unveiled **the world's first electromagnet** in 1825. It was a horseshoe-shaped piece of iron with a coil of copper wire wrapped around it; when a current was passed through the coil it became magnetised and when the current was stopped it was de-magnetised. Sturgeon, from Whittington, Lancashire, proved its power by lifting a 9lb weight with a seven-ounce piece of iron wrapped with wires powered by the current of a single cell battery. This was a very significant breakthrough; as the US Department of Energy's Office of Science says, 'In today's world almost all jobs other than a goat herder use some type of electromagnet... they are everywhere'. The point about the electromagnet was that, unlike a normal magnet, its magnetism could be regulated – switched on and off. This enabled the use of electrical energy to control parts within machines.

James Clerk Maxwell (see *Photography* and *Microwave Ovens)* lived for just 48 years and, despite being virtually unknown today, he was one of the most important scientists of the 19th Century. Born in Edinburgh in 1831, he wrote his first mathematical paper (on the properties of curves) at the age of 14, went to Edinburgh University two years later and was a professor at Aberdeen at just 25. His greatest achievement was to develop a set of equations which, when published in 1873, finally explained how light and electromagnetism were basically different forms of the same thing. (These equations are not easy to understand or explain simply, but they underpin all our modern technology, from power generation to mobile phones and television. Einstein built on them when he developed his Theory of Relativity and described Maxwell's work as 'the most profound and the most fruitful that physics has experienced since the time of Newton'.)

Maxwell showed that electromagnetic fields travel at the speed of light, which he calculated as 310 million kilometres per second, only slightly higher than the currently accepted value of 300 million kps. The lack of wider recognition of his achievements today may be because his equations do not reduce to an easy tag line, such as $E=mc^2$, and there are no easily-remembered stories about apples falling on his head. It was 2008 before a statue of him was erected in his home city.

EMERGENCY CALLS

BRITAIN'S 999 system is **the oldest emergency call service in the world**, and was launched after a serious fire killed five women in London in November 1935. The blaze broke out in a doctor's surgery at 27 Wimpole Street in London. The subsequent inquest heard how the fire brigade had reached the scene too late, after a neighbour's telephone call was not picked up quickly by the operator (in those days, calls were routed through an exchange staffed by technicians who answered the call and then physically connected the caller to the intended destination). It was decided that some way was needed to warn operators of priority calls coming into the exchange, and the memorable number 999 was soon agreed upon. All coin-operated telephone boxes (few households had phones at the time) were changed to make 999 calls free, and at the exchange a flashing light and klaxon alerted operators to the emergency call. (The horn was so loud that telephonists would stuff the horn with a tennis ball to muffle the noise.)

The 999 service was introduced in London on June 30, 1937 and the very first 999 call, made by a Mrs Beard, of Hampstead, led to 24-year-old burglar Thomas Duffy being caught red-handed raiding her house. In the first week 1,336 emergency calls were made – 91 of them were hoaxes or practical jokes. Today 30 million 999 calls are made each year. Unfortunately, nearly half of them are said to be made by hoaxers or idiots: among the latter in recent years, one man asked for help with a crossword, a woman rang to say said she could not reach her TV remote and a further caller complained that his wife had gone out leaving only two salmon sandwiches for his lunch.

THE EMPIRE

IT'S fair to say that few people now regard Empires as wholly good things – but they're not always wholly bad, either. On the one hand, as John Cleese's Judean revolutionary 'Reg' is forced to admit in *Monty Python's Life Of Brian*, the Romans brought aqueducts, sanitation, medicine, education, wine, public order, irrigation, roads, a fresh water system and public health to his country. But on the other, there was the small matter of subjugation and the removal of the right to self-governance and self-determination.

Whichever side of the scale you believe weighs heaviest, it's impossible to deny the *technical* brilliance of the Roman Empire – the sheer fact that it existed over such a large area and lasted for so long is astonishing testament to the skill, courage and determination of the Romans. The same can be said of the British Empire, only much more so. This is not a book about the Empire (Niall Ferguson's *Empire* is a highly recommended in-depth account of its history) but we ought just to reflect on its scale: at its height, in the 1920s, it was **the largest the world has ever seen**, ruling a quarter of the surface of the planet and a quarter of its people. Our largest territory, Canada, covered 3,603,336 square miles. The smallest, Gibraltar, spanned just two square miles. Only 200 people lived on tiny Ascension Island – but Britain was also in control of India, which at that time had a population of 319 million. To many of these countries we gave the rule of law, the basis of democracy, a worldwide language and huge technological advances. We took plenty – from raw materials and slaves to words and foods ('pyjamas' and chicken tikka both came from India, for instance) – and there are arguments to be had about whether our influence was benign or malign, or a mixture of the two. But it's the mere fact that we did it which is astonishing.

It was said that the Sun never set on the British Empire – based on the fact that it was always day-time somewhere we ruled. But the sun finally set on it after the Second World War, when we no longer had the resources to control such huge territories.

ENGLISH

WE were late arrivals to the idea of civilisation. Long after the Egyptians had built their great Pyramids and the Greek, Chinese, Arab and Indian civilisations had risen and fallen, we were still cheerfully painting our faces blue and dancing around naked on the cliffs of Dover, screaming insults at the first Roman visitors. They did not take well to this, unfortunately, and we were forced to do some quick catching up. The efficiency with which we did this is best exemplified by the spread of our language.

The Latin of those Roman invaders is long dead, but **one in five people on this planet now speak English** to some degree, and it's the official language of dozens of countries, from Anguilla to Zimbabwe. The only language spoken by more people is Mandarin

Chinese, and this is simply by virtue of the enormous population of China. English is central to business, science, diplomacy, the internet and air traffic control. It has more words than any other language and although relatively easy to speak, it is fiendishly difficult to learn to write, largely due to the many words which sound the same but are differently spelled – their, there and they're, for instance.

The longest word in the English language is *pneumonoultramicroscopicsilicovolcanoconiosis*, a medical term for a lung disease caused by inhaling volcanic dust; the longest word containing no vowels is *rhythms* and the longest commonly-used word with all five vowels used once in their correct order is *abstemiously*.

THE ENLIGHTENMENT

THE Enlightenment is the name given to the period during which people began to move away from rule by kings, unquestioning acceptance of religion and the use of superstitious beliefs such as alchemy and witchcraft to explain the natural world, towards rationalism and scientific empiricism. It is widely agreed to have begun in the early 1600s, and many philosophers and scientists in many countries were involved. But much of it was driven from Britain: indeed, Thomas Jefferson, third President of the USA (from 1801 to 1809) and the main author of the Declaration of Independence, wrote in 1789 that the 'three greatest men that have ever lived, without any exception' were John Locke, Francis Bacon and Isaac Newton, three English giants of the Enlightenment.

Locke, a philosopher, was born to Puritan parents in Somerset in 1632. He advanced the idea of self-governance and the social contract, by which people allow a government to rule them in return for guarantees of security, and is regarded by some as the father of liberalism and libertarianism. He died in October 1704, and never married.

Bacon was born in 1561 and went to Cambridge University aged 12. He was a leading empiricist, and introduced the Baconian method of scientific reasoning. This was an attempt to explain a given occurrence by deduction from the various factors present in it, as opposed to relying on guesswork or identifying some unknown causal factor. This kind of reasoning seems obvious to us now, but in the 1600s it represented a dramatic insight. A good real-life example would be the discovery that malaria is caused by parasites within mosquitoes, as opposed to the 'bad air' which was originally blamed (see *Malaria*). Bacon was knighted in 1603 and would have been a victim of Guy Fawkes' Gunpowder Plot if it had succeeded. Instead, he became a member of an elite group of scientists who were killed by their own experiments. It occurred to him one winter's day in 1626 that perhaps meat could be preserved by snow. Without further ado, he bought a fowl, stuffed it with snow and got very cold himself in the process. He then caught pneumonia and decided to eat the bird. It wasn't as well preserved as he thought and led to his death shortly after. (We'll look at Newton later.)

FARMING

THE Agricultural Revolution and the Industrial Revolution which followed it were the biggest advances ever seen in human history. The British led them both, developing radical new farming methods which helped to guarantee the food supply and allowed population growth.

Jethro Tull was born in Berkshire in 1674 and studied at Oxford intending to take up a legal career. Due to ill-health, he returned home and later began farming with his father. Agriculture had stood almost still since Roman times: seeds were still scattered by hand in the fields, often to be eaten immediately by birds. Tull realised how wasteful this was and in 1701 he developed a wheeled device called a **seed drill** that was pulled behind a horse and cut three regularly spaced furrows in the ground before dropping seed into them at a constant rate. As the seeds were being planted in neat rows, he was then able to develop a **horse-drawn hoe** which could weed around the lines, increasing the crop yield. These two developments were the first steps towards the mechanisation of agriculture.

Charles 'Turnip' Townshend – in the words of the poet Alexander Pope, he was 'obsessed by turnips' – was also born in 1674, in Raynham Hall, Norfolk. He went into politics and became Secretary of State but he repeatedly quarrelled with his rival and brother-in-law, Robert Walpole, Chancellor of the Exchequer. After one heated argument almost turned to blows in 1730, Townshend resigned and decided to concentrate on farming his estate.

For centuries, a three-field system of crop rotation had been used, where wheat would be grown on a strip of land one year, then barley (for livestock to graze on) the next, followed by a year where it was left empty (fallow) to recover. Townshend pioneered what became known as the **Norfolk Crop Rotation System**, where the order of growing was wheat, clover, turnips then barley. Because the clover put nitrogen back into the soil, no field had to be left fallow. This new system also meant that livestock could be better fed, as they could graze on the turnips and clover. (There is some debate as to whether Townshend actually came up with the idea, or just had the brains to experiment, modify and popularise it before he died of apoplexy in 1738.)

Now that we could feed the cattle, the next thing to do was to try and make them taste better. **Robert Bakewell** was born and brought up in Dishley near Loughborough, and took over the family farm in 1760. People were leaving the countryside and moving into the towns, and this meant that more mouths had to be fed by fewer farmers. Bakewell pioneered the **selective breeding** of sheep and cattle, allowing only the best animals to breed (previously, males and females had been kept in fields together and allowed to breed randomly). The effect of this was to increase the average weight of each animal, and to improve the taste and quality of its meat. As other farmers followed his lead, farm animals increased dramatically in size and quality: in 1700, the average weight of a bull sold for slaughter was 168kg. By 1786, that weight had more than doubled and his methods were being used worldwide. Fifty years after his death in 1795, Bakewell's work would be referred to by Charles Darwin as an example of Artificial Selection.

THE FAX MACHINE

IT won't surprise you to learn that the fax machine is a British invention. It *may* raise your eyebrows to hear that it was invented in 1843.

Alexander Bain (1810-1877) was the son of a Scottish crofter who travelled to London to work as a clockmaker after an undistinguished school career. He proved to be a prolific inventor. Among his designs were for an electric fire alarm and an electric clock. Reportedly, he approached the famous and wealthy Gloucestershire scientist Sir Charles Wheatstone for funds and assistance for the development of the latter, only to be told, 'Oh, I shouldn't bother to develop these things any further… there's no future in them.' Imagine his surprise when Wheatstone demonstrated an electric clock to the Royal Society three months later, claiming it was his own invention. Luckily, Bain had already applied for a patent and Wheatstone was disgraced for his attempted thievery.

Bain's fax machine, which featured a mechanical stylus attached to an electromagnetic pendulum, enabled him to send typed words over a telegraph line (see *Telegraph*). Wakefield-born **Frederick Bakewell** (1800-1869) improved on Bain's design, and demonstrated a facsimile machine which could send pictures in 1851. The present BT building in Glasgow is named after Bain.

FEMALE DOCTORS

ELIZABETH Blackwell (1821-1910), the daughter of a sugar refiner from Bristol, was **the first woman doctor** in both Britain and America.

In 1832, when she was 11, Elizabeth's family moved from the West Country to New York. When her father died suddenly, the teenaged Blackwell became a teacher to earn money for her mother and siblings; she turned to medicine when a friend died from uterine cancer, having been too shy to see a male doctor.

Despite the patent madness of excluding 50 percent of the population, medicine was not considered a suitable profession for women and Blackwell was rejected by 29 medical schools. In 1847, after several years of private study, she applied to join a school in Geneva, New York State. The faculty, believing that the male students would reject her application, allowed them to decide on her admission and, as a joke, they voted unanimously to admit her. The joke promptly turned sour when it was clear she was serious and able, and she was shunned and barred from practical demonstrations. But she had the last laugh when, in January 1849, she graduated top of her class, beating all 150 male students. In so doing, she became the first woman in America to graduate from medical school.

Ten years later, during a year-long lecture tour, Blackwell's was the first woman's name to be added to Britain's General Medical Council Register, entitling her to work as a doctor here also.

In 1857, with her sister Emily, who also became a doctor, she opened the New York Infirmary for Women and Children. It was **the first hospital operated by women** and the first to offer clinical training to women. In 1869, Blackwell returned to England, living in London and then Hastings until her death in 1910. By 1881, there were still only 25 female doctors in England and Wales; now more than 50 percent of medical students in the UK are women.

The western world's first woman doctor had to pose as a man to be accepted. Miranda Barry called herself 'James' and stayed in character for 50 years after graduating from Edinburgh Medical School in 1812. She was a military surgeon and eventually Inspector General of Hospitals, and, says www.scotland.org, a 'notorious dandy and flirt (who once even fought a duel over a woman)'. Her gender was only revealed after her death in 1865.

FEMALE HYSTERIA

IN the late 19th century, a strange medical condition which only affected women spread rapidly through Britain. When one considers the cure, perhaps it's hardly surprising.

Called 'female hysteria', the symptoms ranged from insomnia to rapid heart beat to loss of appetite or faintness – one physician of the era said that up to a quarter of women suffered from it and catalogued a list of 75 symptoms. The treatment? Genital massage, given by the doctor to the patient until she experienced 'hysterical paroxysm'.

It is often said that the past is a foreign country, where they do things rather differently, and the condition is no longer recognised. Indeed, modern doctors attempting to lay hands on their patients are apt to find themselves struck off. Back then, however, some wives were apparently so enthusiastic about the miracle remedy that they made sure to visit their physician at least once a week, at their husbands' expense. Of course, if a quarter of all women are in urgent need of regular medical manipulation, even the most conscientious of physicians will find his fingers aching before long. Thus it came to pass that the puritanical Victorians were behind an unlikely British invention which has since been adopted with some enthusiasm by ladies across the globe: the electric vibrator.

Joseph Mortimer Granville, a British doctor born in 1833, came up with the solution and in 1883 patented the *perceteur*. It was one of the earliest electric devices and it replaced a *steam-powered* version, which conjures up heroic visions more suited to other publications altogether. Granville's 'aid' was soon being advertised in magazines as staid as *Woman's Home Companion* and *Needlecraft*, with the recommendation that it would 'furnish every woman with the very essence of perpetual youth'.

FIGURE SKATING

OK, perhaps it didn't make Britain *great*, but **the first ice skating association** was formed in Edinburgh and the original how-to-do-it book on ice skating was published in London in 1772, having been written by a lieutenant in the Royal Artillery, **Robert Jones**. Later, we undid much of this good work by bringing Jayne Torvill and Christopher Dean to the attention of the world.

FINGERPRINTING

IN JUNE 1892, police were called to a house in Necochea, on Argentina's south Atlantic coast. Inside, they found a terrible scene: two young children had been butchered and their mother, Francisca Rojas, had had her throat cut. She was still alive, however, and named a neighbouring ranch worker called Velasquez as the attacker. He was arrested and questioned, but refused to confess to the murders even under torture.

After nine days, a detective called Juan Vucetich was brought in from Buenos Aires. He ordered that the fingerprints of both the suspect and the mother be taken and a thorough search of the crime scene be carried out. A number of prints in blood were found on a doorpost: they matched those of Francisca Rojas and not the arrested man. Rojas confessed to the murders – it seems Velasquez had rejected her advances and she had been trying to punish him – and was jailed for life. She became the first person to be convicted of murder on fingerprint evidence. And although the detective, Vucetich, was Argentinean, his methods were British.

Fingerprint identification – or 'dactyloscopy', to give it its formal name – is now a cornerstone of modern crime detection. It was, of course, developed in Britain – though several men have competing claims.

Botanist **Nehemiah Grew**, a vicar's son born in Coventry in 1641, was the first to describe patterns on the skin. In 1684, he published *A Theme On Skin Structure*, a paper containing accurate drawings and information about the 'innumerable little ridges, of equal bigness and distance' on the hands and feet.

In 1878, grocer's son **Henry Faulds**, a Scottish doctor and missionary, was at an archaeological dig in Japan when he noticed that fingerprints left by ancient potters were still visible in their work. Shortly afterwards, his hospital was broken into and the police arrested a member of staff. Faulds compared the fingerprints at the crime scene to those of the suspect and showed that they were different. The police were impressed with this new method of identification and released the suspect. This was **the first recorded use of fingerprints in any crime investigation**.

When Faulds (1843-1930) later returned to Britain, he presented his fingerprinting system to Scotland Yard. Incredibly, they turned it down. Jack the Ripper was operating on the Yard's patch around this time: perhaps the world's most infamous serial killer might have been caught if the police had thought differently.

Faulds, who used chemicals to get rid of his own fingerprints only to find that they grew back in exactly the same pattern, had published the first scientific paper on fingerprinting in the science journal *Nature* in 1880. Immediately, the magazine received a letter from a **William Herschel** (See *Uranus*: the planet was discovered by Herschel's grandfather, also named William) claiming that *he* had been using fingerprints for the last 20 years. Slough-born Herschel (1833-1917) was a British civil servant working in India. Part of his job was to ensure that people received their pensions. As most were illiterate, he'd come up with the idea of using fingerprints to identify them.

Unsurprisingly, Faulds was none too pleased about his thunder being stolen and demanded proof. Although Herschel was able to provide this, accusations of plagiarism flowed between them.

Enter the Birmingham inventor and scientist **Sir Francis Galton** (1822-1911). He crystalised the new science, writing three books and an academic paper on the subject in the 1880s. He detailed the eight main patterns (whorls, arches and loops, and sub-divisions thereof) and calculated the chances of a false match at one in 64 billion. The legal system was won over and his classification (Galton's Details) is still used today. (In a twist worthy of any crime drama, it turns out that Faulds had written to Charles Darwin about his ideas many years earlier. It was an attempt to gain credibility for his work, through the support of the eminent scientist. Darwin, however, simply passed the letter on to his cousin... Francis Galton.) Juan Vucetich – the Argentinean detective – had corresponded with Dalton and was able to use his knowledge to nail Francisca Rojas.

Sir Edward Henry (1850-1931), from Middlesex, set up the fingerprint bureau for the Metropolitan Police in 1901. **The first UK conviction using fingerprint evidence** came in September 1902 when one Henry Jackson was sentenced to seven years' penal servitude for stealing billiard balls from a house at Denmark Hill, south London. He had left a thumb print on a newly-painted window sill. Three years later, brothers Alfred and Albert Stratton were hanged for the murders of shopkeepers Thomas and Ann Farrow in Deptford after being convicted on fingerprint evidence. A bloody thumbprint was found on an open cash box.

No two individuals have ever been found to have the same fingerprints – even those of identical twins are different. So there are an astonishing number of distinguishable fingerprints. According to the US Census Bureau, as this sentence is written there are 6,765,880,361 people in the world. Give or take a missing hand or digit here and there, this equals more than 67 billion prints. Visible fingerprints are left on surfaces when our fingertips have blood, grease or dirt on them, and 'latent' prints are formed by the sweat which is always present – the more so if one is engaged in nefarious activities.

THE FIRE BRIGADE

THE world's first municipal fire brigade was set up in Edinburgh in 1824. Led by **James Braidwood**, a buildings surveyor who was born in the city in 1800, Edinburgh Fire Engine Establishment became renowned for its efficiency, and its fit and disciplined men. Braidwood, who led from the front – he once entered a burning building with a wet blanket over his head and removed two barrels of gunpowder from the cellar – insisted that his firemen rise and do drills at 4am. He also employed new fire-fighting methods, including getting as close to the seat of the fire as possible, and invented the chain ladder to enable people trapped in burning buildings to escape.

Although other towns and cities had a fire service in the early 17th century, they were organised by insurance companies who would only put out blazes in their customers' buildings. Word spread about Braidwood's success and, when several London fire insurance companies amalgamated in 1832, he was paid £400 a year (£320,000 today) to set up the London Fire Engine Establishment. His team tackled fires at the Houses of Parliament and Windsor Castle. Braidwood died when a wall collapsed on him as he fought a vast blaze in riverside warehouses in Tooley Street in June 1861. Iron fire doors, which he had recommended be installed, had been left open, rendering them useless. He left a fortune worth £3.5 million today.

THE FIRE EXTINGUISHER

AND if the fire brigade are otherwise engaged... **George Manby**, a soldier and writer from Norfolk who went to school with Lord Nelson, invented **the world's first portable fire extinguisher** in 1816 (though some sources say 1813 or 1818). His '*Extincteur*' consisted of a four-gallon copper cylinder which held three gallons of potassium carbonate dissolved in water and a quantity of compressed air. When required. the water was forced out through a tube and directed toward a fire.

Manby's interest in life-saving started in 1807 when he witnessed the *Snipe* disaster in the harbour at Great Yarmouth. The ship ran aground in bad weather on rocks just off shore, and 67 men, women and children drowned as onlookers watched in helpless horror from less than 100 yards away. Manby (1765-1854), who had served in the

Fusiliers, adapted a cannon to fire a rope from shore to ship. Sailors would shin up the mast, secure the rope and use it to slide to safety. It took some development – the rope needed to travel in a straight line, without burning or tangling – but the first 'Manby Mortar' was used in 1808 to bring seven crew off the sinking prison ship *Elizabeth*. When the Preventative Water Guard was established the next year (the precursor to HM Coastguard), each station was issued with a Manby Mortar.

FIZZY DRINKS

FORGET Coca Cola: **the world's first fizzy soft drink** was made by an English vicar. **Dr Joseph Priestly** (1733-1804), a clergyman and chemist from Birstall, near Leeds, is best remembered by scientists because he isolated and identified seven gases, including oxygen. But he also invented soda water in 1772. Priestley lived next door to a brewery and noticed the gas coming from the fermenting vats. He recreated the gas, carbon dioxide, at home and called it 'fixed air'. He noted that it made water bubbly and pleasant to drink. His carbonated soda water was a huge hit.

In 1790, Jacob Schweppe, a Swiss watchmaker and amateur scientist, set up a factory in London in 1790 to apply Priestley's invention commercially. Schweppe was the first person to come up with a reliable machine to add fizz to water (which he sold as a cure for indigestion and gout). It wasn't until the 1830s that bubbles were added to other drinks, such as lemonade. Schweppes Tonic Water was introduced in the 1870s.

THE FLUSHING TOILET

QUEEN Elizabeth I was the proud owner of **the world's first flushing loo**. It was made for her by her godson, **Sir John Harington** (1560-1612), who called his invention 'Ajax' – a pun on 'jakes', a colloquial Tudor name for a lavatory. According to his biographer Jason Scott-Warren, the Queen banished Harington from court in 1584 for regaling her ladies-in-waiting with a risqué Italian poem. She told him not to come back until he had translated all 33,000 lines of the work into English.

During his exile, he translated the work, dedicating it to Her Majesty. And in an extra effort to get back in her good books he also came up with the radical idea of placing a reservoir of water above a privy with a flushing mechanism to get rid of waste. The pan had an opening at the bottom which was sealed over until the user pulled a system of handles and levers, pouring water into the pan and opening the valve. He installed one Ajax in his home in Kelston, near Bath, and another in Richmond Palace for the Queen. It never caught on and people continued to use chamber pots, despite Harington's 1596 work *A New Discourse of a Stale Subject called The Metamorphosis of Ajax*.

More than 200 years later, in 1775, **the world's first patent for a flushing toilet** was issued to Scottish watchmaker and mathematician, **Alexander Cummings**. Born in Edinburgh in 1732, Cummings was working in London when he came up with a method of keeping water in the bowl to stop foul sewer smells seeping back. Three years later, his design was improved upon by **Joseph Bramah**. A cabinetmaker born in Stainborough, Yorkshire, in 1748, he used a plunger to release the waste and seal the soil pipe, and also designed a float valve system for the cistern. In the mid 1850s – not that we were obsessed, or anything – another Englishman, **Josiah Jennings**, from Eling, Hampshire, made further improvements and invented **the world's first public toilets**. They were installed at the Great Exhibition of 1851 and caused great excitement. Called Retiring Rooms, 827,280 visitors paid one penny to use them – hence the phrase 'to spend a penny'.

Thomas Crapper, the unfortunately-named plumber often wrongly credited with inventing the flushing toilet, was born in 1837 in Thorne, Yorkshire. He held patents for nine plumbing innovations, including the floating ballcock, but his firm did not make toilets, they merely fitted them.

FOOTBALL

FOOTBALL is the world's No1 sport. Billions play or watch it across the globe, and it has been estimated, admittedly by me, that there are only 14 living people, a family of nomads in the Kalahari, who have not heard of David Beckham. (Lucky them, eh?)

Unsurprisingly, the Han Dynasty Chinese had a 'similar' thing called Tsu Chu, which involved kicking a feather-filled bag through a 2ft-wide 'goal'. And the Greeks played 'Episkyros', and the Romans played 'Harpastum', and the Ancient Egyptians aimlessly hoofed pebbles around pyramids, and the Mayans…

No. No. *No*. The undisputed king of global games is a British invention, and everyone else is just going to have to get used to that. FIFA, the world governing body, says quite plainly, '**It all began in 1863 in England**.' Bear in mind that 'FIFA' stands for the 'Fédération Internationale de Football Association'. Given how firmly it sticks in the craw of our European cousins to give Great Britain the credit for *anything*, we can probably take it that they have explored every possible alternative origin before, grudgingly, admitting the truth.

According to footballnetwork.org, the game flourished in England from the 8th century, mostly as a sort of violent free-for-all involving up to 1,000 people. 'Players were kicked and punched regularly by opponents,' it says. 'Damage also occurred to people's houses and businesses.' So not much different to the modern game, then, though nowadays the players mostly leave this sort of behaviour to the supporters (Britain, mainly England, having regrettably also popularised football hooliganism).

Football actually sprang out of England's poshest schools. In 1848, the 'Cambridge Rules', an early attempt to formalise the game, were drawn up at the university of that name at a meeting attended by representatives from Eton, Harrow, Winchester, Shrewsbury and Rugby. The 'Sheffield Football Club Rules' followed a decade or so later and they, too, were drawn up by ex public schoolboys. (It seems quite likely that ex public schoolboys were the only people in Sheffield with enough time on their hands to bother attempting to codify the kicking around of an inflated pigs's bladder. Most other men in Sheffield at the time were busy working themselves to death in the pits or dying aged 40 of consumption.)

Finally, in 1863, the Football Association was set up at a pub in Great Queen Street, London. Its birth wasn't without controversy. One rule which was originally proposed but later withdrawn would have allowed players to run with the ball in hand; the decision to outlaw this led to a number of leading football clubs pulling out of the FA and instead forming the Rugby Football Union. Another proposed but withdrawn rule allowed the hacking down of opponents at their shins; some modern-day players appear unaware that this was not approved.

The rules were finally standardised in 1882 by the International Football Association Board, which was less 'international' than it sounded, in that it comprised the English, Welsh, Scottish and Irish Football Associations – ie, it was British with a little help from across the Irish Sea. In 1904, the rest of the world woke up, and FIFA was formed. Shortly afterwards, it adopted the British rules under which everyone now plays (with a few revisions).

Britain boasts **the world's oldest football competition (the FA Cup) and league**, and the first official international match took place in 1872 between Scotland and England in Glasgow. But typically, despite creating the game, we've rarely been much good at it. Scotland, Northern Ireland and Wales have never won any major competition and sadly almost certainly never will; England has just one World Cup triumph, in 1966, to its name, amid an otherwise endless sea of mediocre performances.

THE FRIDGE

FOR years, the only way to chill food (or anything else) was to collect huge piles of ice every winter and store it in cold rooms underground until it melted – hopefully not before the following winter came along to replenish supplies. Then – almost 300 years ago – a British scientist pioneered **artificial refrigeration** (when he discovered the principle of heat exchange). This is where heat from one substance is absorbed by another, making the first warmer and the second cooler – and it's the basis upon which your kitchen fridge works.

It was first demonstrated in Britain by **William Cullen** in 1748. Cullen (1710-1790), an Edinburgh doctor who was born in Hamilton, near Glasgow, was interested in chemistry. He performed the world's first experiment in artificial refrigeration at the University of Glasgow with a container of diethyl ether, a dangerously flammable substance made by adding sulphuric acid to alcohol. Cullen pumped the air out of the container, creating a vacuum and making the ether boil. (The boiling point of a liquid falls as the air pressure around it decreases. This is why you can't make a decent cup of tea on Mount Everest. Tea requires water at 100°C, but it boils at a lower temperature at higher altitudes where air pressure is lower.) In the process of boiling, the ether absorbed heat from its surroundings, causing water vapour in the air to form ice crystals on the outside of the container as the temperature dropped.

Interestingly, people were still being executed for 'witchcraft' in Scotland up to 1727, and the penalty only changed to imprisonment in 1763; Cullen was treading a fine line, like so many other early scientists. His discovery did not signal the end for the ice house, though, as he was not able to develop his idea commercially. Later on, others did – mostly in America. The first household refrigerators, using Cullen's basic principles of heat exchange, were introduced in 1911 and sold for an incredible $1,000 ($22,512.70 today). A Cambridge company, Camfridge, is currently working on revolutionary technology using advanced metals, nano technology and magnetism to create the fridges of the future.

THE GAS MASK

EDWARD Harrison, inventor of **the first practical gas mask**, was a brilliant chemist and a true British hero. He tried to join the Army at the outbreak of WWI, but was rejected as too old (he was 45, having been born in Camberwell, London, in 1869). Undeterred, he reapplied and was accepted into the 'Sportsmen's Battalion' which recruited older soldiers. However, when the Germans used gas for the first time, any chemists in the ranks were quickly reassigned to work on ways of defeating this terrible new weapon.

He was promoted to Lieutenant-Colonel and placed in charge of the team which eventually created the gas mask. The Royal Society of Chemistry records that Harrison 'undermined his own health so badly that he died a week before the end of the war on November 4, 1918, having worked night and day, against medical advice, to achieve his goal of protecting millions of men from agonising death and disability'. Winston Churchill, then munitions minister, wrote a letter of 'condolence and admiration' to Harrison's wife, Edith – who also lost her eldest son in the war, at the Somme in 1916.

GIN AND TONIC

THE world's most civilised cocktail was created by British soldiers in India in the 1820s. Tonic contained quinine – a bitter-tasting malaria preventative – and gin was used to sweeten it. Slices of lemon prevented scurvy (see *Malaria*).

GOLF

THE story goes that 12th century Scottish shepherds passed the time by knocking stones into rabbit holes on the Fife heaths where St Andrews Golf Club now sits. This sounds far-fetched – how could it ever have caught on? – but there's no doubt that the rules and arcane rituals of this great sport as played today developed on these islands. **The oldest golf course in the world** is Musselburgh's Old Links, where golf was played by Mary, Queen of Scots (and others) as long ago as 1567 (though the game had been played for many years before that). St Andrews laid down the standard number of holes – 18 – in 1764, and the Royal and Ancient Golf Club of St Andrews (set up in 1754 by 22 noblemen and gentlemen) is the international guardian of the rules.

GRAY'S ANATOMY

THE world's first anatomical textbook, originally called *Gray's Anatomy: Descriptive and Surgical Theory*, was created by London doctor **Henry Gray** and produced with his colleague **Henry Vandyke Carter**, from Hull, in 1858. A stunning book full of highly detailed drawings of body parts and descriptive text, it is still the leading authority on the subject 150 years later (though it has frequently been revised in that time). The two men spent 18 months working on the first edition, dissecting unclaimed bodies from hospital and workhouse mortuaries. Gray, born in 1827, died of smallpox at the age of just 34. He contracted the disease from his dying nephew while studying its anatomical effects on the human body.

GREAT ANTARCTIC ENDEAVOURS (1)

WHEN the British go exploring, we like to do it with as much suffering as possible – often failing to make it back at all and instead dying tragically along the way.

If this sounds incongruous in a book primarily about Great British triumphs (or, at least, successes), it really isn't. In striving for greatness, it is often necessary to take great risks – and in taking great risks, even great people can fail, sometimes disastrously. And if you're going to fail disastrously, there are few places better suited for the task than the frozen, empty wastes of the Antarctic, as **Robert Falcon Scott** and his men discovered to their cost.

Antarctica is a huge continent – at well over five million square miles, Great Britain would fit into it 50 times – surrounded by the vast Southern Ocean. Roughly in the centre is the South Pole – the southernmost point on the surface of the globe. There is no life at the Pole, or for many miles in any direction, which isn't surprising given that 98 percent of the land mass is covered by ice, to an average depth of one mile. It is high and exposed, very windy and extraordinarily cold: the lowest temperature ever recorded on earth was a reading of −89.2°C taken at Russia's Vostok Station in July 1983.

Scott, a charismatic Naval officer born in Devon in 1868, dreamed of becoming the first man to stand on the South Pole. But it was a dream which became a nightmare, as he was beaten by the Norwegian Roald Amundsen and died horribly with four others.

He and his expedition of 33 souls arrived at Cape Evans on the Antarctic Coast to prepare for their attempt in January 1911. His plan was for a party of five, including himself, to reach the Pole. They would haul as much food and other supplies on sledges as possible, but the sheer distances involved – the return journey was 1,766 miles – meant that supply depots would also need to be laid along the way beforehand. During the first few months after their arrival, a series of depots were put down at average intervals of 65 miles from the coast to the Beardmore Glacier which stood halfway to the Pole. This was finished by the end of April, and they settled down to spend the dark, cold winter (the northern hemisphere summer) in their hut.

Perhaps he was bored, but Dr Edward Wilson, a veteran Antarctic explorer from Cheltenham, suggested a 60-mile trek over the ice to Cape Crozier to collect Emperor Penguin eggs which he believed would shed light on the evolutionary link between birds and reptiles. That wouldn't tempt me from my sleeping bag, but two men volunteered immediately: Henry 'Birdie' Bowers, a Naval Lieutenant, and Apsley Cherry-Garrard, who had no previous experience, was extremely short-sighted and had paid £1,000 to join the expedition as 'Assistant Zoologist'. That 19-day journey alone was an epic feat, as temperatures dropped below -60°C and clothes and bedding froze solid. Just when it seemed that things couldn't get any worse, their tent blew away in a screaming blizzard. Knowing that they were dead if they could not recover the tent, the trio searched desperately and miraculously found it caught around a large rock. As a final kick in the teeth, only three eggs made it back intact and none of them proved Wilson's theories. (It's all contained in Cherry-Garrard's book, *The Worst Journey In The World*, a work of classic British understatement.)

On November 1, 1911, after the daylight returned, the expedition proper began. An initial party of 16 set out, using ponies and dogs to pull their sledges until they reached the Beardmore Glacier. At that point, the ponies were shot for meat and the party was divided, with some of the men returning to base camp to await Scott's return. (They took the dogs with them since Scott believed that using dogs to pull the sledges was akin to cheating. In contrast, Amundsen's Norwegian expedition, which had set out for the Pole on October 19, was using dogs all the way.)

Scott and his remaining men, now hauling their own sledges, often up almost-vertical slopes, pressed on. Seven more turned back at two predetermined positions, having set up more supply depots. That left the final five alone on January 4, 1912, facing a journey of 150 or so miles to the Pole.

They were Scott himself, Wilson, Bowers, Army Captain Lawrence 'Titus' Oates and naval Petty Officer Edgar Evans. Evans was a giant Welshman and a father of three; Bowers a 5ft 4in redheaded Scotsman of 28 years known for his toughness and cheerful disposition; Wilson a 39-year-old affectionately nicknamed 'Uncle Bill' by his colleagues; and Oates a Boer War veteran from London with one leg an inch shorter than the other after a gunshot wound.

They set off and, after 11 bone-achingly arduous days, they reached the Pole on January 17 – only to find that Amundsen had beaten them to it. His flag was standing proudly in the ice, together with some supplies and a note to Scott.

Scott's diary records his anguish at this crushing disappointment: 'The worst has happened,' he wrote. 'All the day dreams must go… Great God, this is an awful place.'

The five men turned around for the 800-mile slog back to the coast, but the return journey turned into a disaster. Initially, they made reasonable progress, averaging 14 miles a day for three weeks, but the weather was unusually cold that year and the men grew steadily weaker. Exhausted from the constant hard work of pulling their sledges, and ravaged by frostbite, they were also terribly malnourished – the calorific value of their rations was not high enough and vitamins had only recently been discovered and were not really understood at the time.

The first casualty was 35-year-old Edgar Evans. After suffering head injuries in a fall, he died on the Beardmore Glacier on February 17.

The weather worsened, and the pace of the remaining men dropped further as they struggled on through the seemingly never-ending Antarctic landscape. Titus Oates began to suffer more and more, and was holding the others back and consuming precious food in doing so. One night, as they lay shivering in their tent, Oates, 31, stood up to leave. His immortal words of self-sacrifice, recorded in Scott's diary, have echoed down the ages: 'I am just going outside and may be some time.' His colleagues would have been under no illusion that he was laying down his life to try and save theirs.

Scott, Bowers and Wilson laboured desperately on, but a savage blizzard pinned them down on March 20. Just 11 miles from 'One Ton' food depot, they pitched tent, hoping to sit out the storm. But their supplies went before the snow did. Nine days later, on March 29, Scott – 43, and married with a son – made his final diary entry: 'We shall stick it out to the end, but we are getting weaker, of course, and the end cannot be far. It seems a pity, but I do not think I can write any more. R. Scott. For God's sake look after our people.'

Imagine their torment as they lay helplessly in their tent, weak beyond description, waiting for death with the wind howling mercilessly outside the canvas: Scott has had his share of armchair critics, but few have risked their own necks in the way he did. Although they failed, he and his men were true British heroes.

GREAT ANTARCTIC ENDEAVOURS (2)

SOME people might have taken fright at Scott's fate and given the Antarctic a miss for a while, but Ernest Shackleton wasn't the type to take fright at anything. Although actually born in Ireland in 1874, he was of English descent, he grew up in Kent and his missions were British.

In his *Nimrod* expedition of 1908/9, he had himself come close to dying down there – he and his men had got to within 100 miles of the South Pole before ration shortages forced him to turn back. Even then, they almost starved on the return journey.

Once Amundsen and Scott had reached the Pole, Shackleton saw a crossing of the continent on foot as the last great challenge it had to offer.

Legend has it that he placed the following advertisement in a newspaper: 'Men wanted: for hazardous journey. Small wages, bitter cold, long months of complete darkness, constant danger, safe return doubtful. Honour and recognition in case of success.' Not many people's dream job, but he had no shortage of volunteers and his 'Imperial Trans-Antarctic Expedition' left Britain in August 1914. It consisted of two ships which would land at either end of the land mass. Shackleton and five other men of the 28 aboard his ship, *Endurance*, would make the crossing from north to south; the others were to carry out observations and experiments.

Things went badly wrong in early February 1915, when *Endurance* became trapped in the sea ice just off the coast. Shackleton hoped that she would be freed in the thaw, and ordered his men to sit tight through the long, dark winter as they drifted 1,000 miles east with the ice. Unfortunately, in late October, the ship was crushed by the enormous, shifting slabs pinned against her sides. Supplies, sled dogs and three lifeboats were hurriedly transferred to the floes as she was abandoned to her fate. All hopes and plans for the crossing were now forgotten – stranded on the sea ice, with no means of contacting the outside world, the mere survival of the men was now in doubt.

Initially, they hoped to walk across the ice to the open sea, so that they could row to one of several islands where earlier expeditions had been based and where they knew emergency food supplies were located. But the movement of the water and ice threw up huge, 15ft ridges in the surface – dragging their food, fuel and heavy lifeboats over these was impossible. On December 29, 1915, Shackleton decided that their best hope lay in camping on the flattest, biggest section of ice they could find, and waiting for the summer thaw to bring the sea to them. They called this 'Patience Camp'.

As the weeks turned into months, Patience Camp and the men on it drifted hundreds of miles north-east. They were cold, fearful and soon close to starvation, and supplemented their rations by eating penguins and seals when they could shoot them. By March 30, 1916, Shackleton ordered that their dogs be shot and eaten. Then, on April 8, Patience Camp suddenly split in two, leaving the party floating on a perilously small floe in the freezing sea. They took to the lifeboats and Shackleton struck out for the nearest land. This was Elephant Island, a barren, uninhabited and thoroughly inhospitable spit of rock, and the journey there was hellish: the crew were tired, hungry and desperate, there were icebergs all around and they were soaked to the skin by saltwater in temperatures as low as -30°C. They made it after several days but, in its desolate isolation, far from shipping channels, the island could offer only a relatively brief respite against a slow death from starvation in the bitter cold of the Antarctic.

Lesser men may have accepted their fate at this point, but Shackleton was not, of course, a lesser man.

On *Endurance*'s journey south, she had called in at South Georgia, an island to the north. They had spent time with Norwegian whalers there: Shackleton began drawing up plans to sail with a few of the men to South Georgia, from where a rescue could be mounted.

Put like that – 'to sail to South Georgia' – it doesn't sound too bad. But this was a journey of 900 miles across the wintry South Atlantic ocean, where roaring gales whipped up freezing seas to 60 feet and more, and where an error of navigation meant they would miss their pinprick of a target and perish. Even if they landed, they would have to slog their way through the mountainous, glacial and uncharted interior of the island to reach the whalers.

Shackleton had the *Endurance*'s carpenter, Harry McNish, modify one of the 22ft lifeboats as best he could – the wooden hull was caulked with seal fat, makeshift sail masts were added and the sides raised – and it was renamed the *James Caird* in honour of one of the expedition's sponsors. On April 24, with a crew of five and enough supplies for four weeks – if he hadn't reached South Georgia within that time, they were dead anyway – Shackleton set off, leaving his second-in-command, Frank Wild, in charge of those left behind.

It was a terrible, two-week voyage. The men were soaked through by spray thrown up by the Force Nine winds, and almost every waking moment (sleep was virtually impossible) was spent chipping away ice forming on the masts to prevent a capsize. The navigator, *Endurance* Captain Frank Worsley, had to wait for odd glimpses of the sun, in order that he could plot the position and course of the rolling, pitching boat. But eventually, against monumental odds, they reached their destination. Leaving some of his exhausted men to recover, Shackleton set out for the whaling station at Stromness with two others. This involved hiking across the island, itself a very significant feat in the circumstances. It took 36 hours – several modern and well-equipped professional mountaineering groups have walked the same route, but none have done it as fast. Eventually, they reached the station, where, haggard and gaunt and suffering from exposure and frostbite, they were greeted by the open-mouthed Norwegians.

After resting up for a short while, Shackleton made several attempts to break through the sea ice to return to the larger party left behind on Elephant Island. Eventually he made it, on August 30, 1916 – more than two years after the expedition had left Plymouth. 'I felt jolly near blubbing for a bit and could not speak for several minutes,' wrote Frank Wild in his diary, of the moment when Shackleton hove into view. So ended perhaps the most astonishing story of human survival ever written.

Stunning photography from the expedition, including of the *Endurance* trapped in the ice, of the launching of the *James Caird* and of the stranded men cheering on Elephant Island as the rescue boat appears, can been seen at http://www.coolantarctica.com – click on 'History' then 'Shackleton: Endurance'.

Not content with nearly dying in the Antarctic twice, Shackleton actually did so in 1922, suffering a heart attack at South Georgia on a later expedition.

It's not all failure, of course (even if most of it is). In 1958, **Antarctica was finally crossed for the first time** by a British expedition led by **Sir Vivian Fuchs**, a geologist from the Isle of Wight. His team travelled across the continent in 99 days using Sno-Cats and specially-adapted tractors.

Ranulph Fiennes (actually, 'Sir Ranulph Twisleton-Wykeham-Fiennes, 3rd Baronet') and **Dr Mike Stroud** (just 'Dr Mike Stroud') achieved **the first *unsupported* crossing on foot** in 1993. Stroud is a doctor from London whose speciality is researching human endurance under extreme conditions, often using himself as a guinea pig. Fiennes had been thrown out of the SAS for blowing up a film studio. He had led the Transglobe Expedition which circumnavigated the globe via both Poles between 1977 and 1979 and these days spends his time climbing the North Face of the Eiger between heart attacks, despite having lost half his fingers and toes to frostbite.

GREAT ARCTIC ENDEAVOURS

ANTARCTICA is not the only frozen wilderness in which ludicrously brave folk with a love of the bracing outdoor life and a taste for impossible odds can tweak the nose of death. The opposite end of the earth is no different.

Unlike the Antarctic, the Arctic Circle is not a continent but describes a huge area including the Arctic Ocean and northern parts of a number of countries – Sweden, Norway, Finland, Russia, Canada, the United States (Alaska), Denmark (Greenland) and Iceland. The North Pole itself is at the centre of this circle, and is actually in the middle of the sea – albeit a sea frozen solid by temperatures that fall as low as -50°C in mid-winter (it's warmer than the South Pole because it's at sea level, rather than altitude, and the sea itself retains warmth).

Britain's **Wally Herbert** became **the first man undisputedly to walk to the North Pole** in 1969. (An earlier claim, by the American Robert Peary, is widely discounted.)

Herbert was a York-born explorer who cut his teeth in the *Ant*arctic – though he'd only found himself there after a newspaper fell on his head from a luggage rack in a bus and opened on a page offering a vacancy for an expedition. After sledding thousands of miles across the South Pole on surveying missions, he decided to head north and, aged 34, led an expedition of four men and 40 dogs on a huge journey of over 3,500 miles across the frozen Arctic Ocean from Alaska in the west to Spitsbergen, a Norwegian island in the east.

He had spent time beforehand training and living with the Inuit; they laughed at his plan to cross the Arctic ice-cap, and marked his map with various points at which they thought he would die.

Herbert and his companions set out on February 21, 1968, reached the Pole on April 6, 1969 and finally got to Spitsbergen on May 29. The reason it took so long – apart from the huge distances involved – was that it was not possible to cross in one year. Much of the ice melts over the summer, so the team camped on an iceberg between July and October 1968 before setting off again. Equipment and supplies – including 27 kg of pipe tobacco and 6,200 cigarettes – were dropped to them by Royal Canadian Air Force planes along the way, making it a less dangerous and arduous task than those undertaken by Scott and Shackleton – though only slightly. (For one thing, they had polar bears to contend with.)

The epic slog, for which Herbert was knighted, has never been repeated, and has been described as 'The Last Great Journey on Earth'; the then Prime Minister Harold Wilson called it 'a feat of endurance and courage which ranks with any in polar history'.

In 2003, **Rupert 'Pen' Hadow**, from Perth, Scotland, became **the first and only man** to trek the 478 miles from the Canadian coast to the North Pole **solo, unsupported and without resupply**. It took 64 days, pulling a 19-stone sledge. Two thirds of the way there he fell through thin ice, almost drowning, getting hypothermia and losing a ski. As if that were not enough, he spent half the time fighting off polar bears – he hit one on the head with a saucepan and it ran off. Hadow had been prepared for all of this since childhood, when his parents hired a nanny, Enid Wigley, specifically to toughen him up. Enid, who had looked after Scott of the Antarctic's son Peter, dressed him in a vest and shorts and made play him outdoors in all weathers while regaling him with tales of polar exploration.

The *first* Britons to travel to the Arctic, however, had done so hundreds of years earlier. Ever since the discovery of America, people had dreamed of finding the fabled 'Northwest Passage', a sea route linking the Pacific and Atlantic oceans through the islands in the Arctic circle, north of what we now call Canada. This would create a new trading route from Europe to India and China, a more direct way than sailing down down to and around the tip of Africa.

Between June 1576 and June 1578, **Martin Frobisher** – an eccentrically brave sailor and pirate from Wakefield in Yorkshire – led three voyages to find it. The danger of these early expeditions, and the almost suicidal courage of the crews who sailed off into the literal unknown, in tiny wooden sailing ships with basic navigation techniques, is obvious. On Frobisher's first journey, two of his three ships sank on the 72-day journey to Baffin Island, off the east coast of Canada. Five of his men were then promptly kidnapped by native Inuits, and never seen again.

Frobisher returned home with no sign of the Passage and nothing to show for his journey but a cargo of black rocks which he thought contained gold. Despite his subsequent visits, he never found the Passage, or his missing men, though he did bring back hundreds of tons of that 'gold ore'. Unfortunately, it was entirely lacking in gold and was completely worthless – ending his Arctic explorations. Having been knighted for his part in defeating the Spanish Armada of 1588, he died off the coast of Brest in 1594 of a gunshot wound.

By 1776, Parliament was offering a prize of £20,000 (around £25 million today) to anyone who could find the Passage. Captain James Cook searched without success, and the Napoleonic Wars of the early 19th century brought our interest in the Arctic generally to a temporary halt. But in 1819 **John Franklin**, a naval officer from Louth in Lincolnshire, set off to explore the land around the Coppermine River, which empties from the Northwest Territories into the Arctic Ocean. The trip turned into a disaster; only nine men out of 20 returned, having survived by eating lichen and their spare boots (and even, say some sources, each other).

After that, who would ever return? Well, Franklin, obviously. Twenty six years later, despite being 59 and retired, he agreed to lead another trip to the Canadian Arctic in search of the Passage. In May 1845, he left with 129 men on two state-of-the-art steam ships, the *Erebus* and the *Terror*. After being spotted by whalers in Baffin Bay two months later, they were never seen alive again. Now the Government offered a *further* £20,000 reward to anyone who could find them; over the next 10 years, more than 30 groups set out... losing more men in the process than had vanished on the original Franklin Expedition.

In 1854, an Edinburgh explorer called **John Rae** met Inuit tribesmen who showed him silver cutlery from the Franklin party and said that, after losing their ships to ice, the crew had made it to King William Island, one of many scattered around the Canadian Arctic archipelago, where many of them had died lingering deaths from sickness and starvation. Parliament awarded Rae half the £10,000 reward, though there was much disquiet at his report, which suggested that (once again) the survivors had resorted to cannibalism.

Four years later, another party found a message from the Franklin Expedition under a cairn on King William Island. Dated 1848, it confirmed that the ships were trapped in the ice and said that 24 of the crew were dead and the survivors planned to head for the Back River, hoping to reach a trading outpost called Dettah on the Great Slave Lake. This would have involved an horrific march of perhaps 1,000 miles. A further 20 years on, the remains of those who had attempted it were found, still several hundred miles short of their destination. Recent autopsies indicated death by pneumonia or tuberculosis, with lead poisoning (from the new-fangled tinned food aboard ship) playing a part.

The Northwest Passage was eventually first navigated by Norway's Roald Amundsen in 1903–1906, though he paid tribute to the British explorers who had gone before him. But the British yachtsman **David Scott Cowper** became **the first man to circumnavigate the globe via the Passage** in a three year voyage which ended in 1990. If the sea ice around Canada retreats with global warming, as many scientists predict it will, a regular route from the Bering Strait through the archipelago of islands to the north east of Canada will become a reality.

GREAT EXPLORERS – SIR FRANCIS DRAKE

PIRATES are unlikely national heroes – but without Sir Francis Drake, 16th century Tudor England would never have been so prosperous and Spanish could well have become our native tongue. During the reign of Queen Elizabeth I, epic voyages of discovery were being undertaken, and huge profits made by colonising the new world and trading with other countries. But it was a risky business undertaken by only the boldest. Drake, a farmer's son born in 1540 in Tavistock, Devon, was one such, earning his place in the history books for his brilliant seamanship, his ruthlessness in piracy and war and for becoming the first Englishman to sail around the world.

Drake was the eldest of 12 sons, born into a humble family. He went to sea aged 12, and was soon taking part in profitable slave-trading missions. From those he progressed to piracy and, at the end of 1577, he commanded a galleon called *The Golden Hind* on an astonishing voyage around the world. Setting out in December on a three-year journey with five small ships and 164 men, he sailed west and – much to the annoyance of the Spanish – down the coast of South America and through the Spanish-controlled Strait of Magellan between Chile and Argentina. On the way, he raided their ports in Cuba and Peru. During his expedition, he also claimed California for the Queen and fixed it so that the English would run the clove trade in the Spice Islands off Indonesia.

He arrived back into Plymouth Harbour on September 26, 1580, with his other ships sunk and half his crew dead, but with a vast fortune in stolen Spanish gold in his holds: Queen Elizabeth I's own share came to almost £160,000 – something like £4.5 billion today, and enough for the Queen to pay off her entire foreign debt and *still* have £40,000 left over. While not wishing to condone piracy on the high seas, it's almost a shame that we don't have anyone like Drake around now.

The Queen knighted him, he became an MP and bought himself a huge manor house. But his exploits had not gone down well in Spain – where he was known as 'The Dragon' – and in 1585 hostilities between our two countries broke out. Never keen on landlubbering, Drake was back in his boat quicker than you could say, 'Como consecuencia de los constantes ataques de los corsarios ingleses a la flota española y del apoyo inglés a las Provincias Unidas de los Países Bajos.' He headed back to the New World and promptly sacked the Spanish ports of Santo Domingo and Cartagena, and on his way back he captured the fort of San Augustín in Florida for good measure.

This only served to enrage King Philip II of Spain (who had actually been sort of the king of England, as the husband of Queen Mary I, from 1554 to 1558). Philip put a bounty of 20,000 gold pieces on Drake's head, and began planning a full-scale invasion of England with the intention of restoring Catholicism to our barbarous protestant land. His temper would not have improved when, in 1587, Drake sailed a fleet right into the heart of two Spanish ports, Cádiz and La Coruña, occupying them and destroying 37 ships. The audacious attack put back the Spanish invasion by a year. But in 1588, Philip's Armada was ready. Comprising around 130 ships armed with 2,500 guns and crewed by 8,000 sailors, and carrying 18,000 soldiers, the fleet would have been a terrifying sight. Unfortunately for them, Sir Francis – by then vice admiral of the British fleet – didn't terrify easily. Though the famous legend, that he played bowls in Plymouth as the enemy ships approached, may not be true, what *is* clear is that, in several sea battles, he attacked the Spaniards with a vim and vigour which destroyed all Philip's hopes and left thousands of his men dead.

In 1596 Drake, who had married twice but had no children, died doing what he did best – looting. His vessel was off the coast of Panama when he succumbed to dysentery. He was buried at sea in a lead coffin, to the sound of distant Spanish rejoicing.

GREAT EXPLORERS – CAPTAIN JAMES COOK

JAMES Cook became entranced by the sea while gazing out of a grocer's window in the fishing village of Staithes, north Yorkshire. He was 17, the year was 1745 and it was the start of a lifelong passion for sailing which led to the farm labourer's son redrawing the map of the world. From those unremarkable beginnings, Cook became one of Britain's greatest explorers, discovering the east coast of Australia, the Hawaiian Islands and mapping Newfoundland and New Zealand.

Born in 1728 in Marton, Teeside, he became a coal shipper's apprentice in Whitby, learning about navigation and seamanship in the violent seas off the north east coast and teaching himself astronomy, algebra, geometry and trigonometry. In 1755, he joined the Royal Navy as Able Seaman Cook and quickly showed an aptitude for cartography. His skills were put to good use in the 1760s when, during five gruelling voyages, he produced the first accurate maps of the coasts and islands off Newfoundland, Canada.

In 1768, with the rank of Lieutenant, he was placed in charge of the first scientific expedition undertaken by the Royal Society, aboard a former Whitby coal-hauling ship renamed His Majesty's Bark *Endeavour*. The bark (a three-masted vessel) was fitted out with 10 guns and carried livestock for milk and meat; at just 100 feet in length, conditions aboard were terribly cramped (and no doubt smelly) for the 94 scientists, assistants and crew who sailed in her.

They left Plymouth on August 26, headed for Tahiti (itself discovered by Cornish sea captain Samuel Wallis the previous June) to undertake astronomical work and collect samples.

Cook was also secretly charged with a further mission – to find the fabled southern continent, *Terra Australis Incognita*, or 'unknown southern land'. Although Australia itself was known – Dutch explorers had discovered the western coast in 1606 and spent much of the following century charting the western and northern coastlines of 'New Holland' – it was felt that there must still be a further continent somewhere in the vast Pacific and Indian oceans, to balance out the landmass in the northern hemisphere of the globe.

In Tahiti, the locals showed the crew how to prick their skin and dye it, starting a fashion for sailors to have tattoos, while the scientists collected samples and carried out their astronomy. Cook then went in search of the new land. He found various Pacific Islands and circumnavigated and mapped the north and the south islands of New Zealand and claimed them for Britain. (The Maori were amazed by Cook's boat, at first thinking it was a floating island or giant bird.) He sailed on, planning to hit Tasmania (discovered by the Dutch) and see if it was part of this unknown continent, but the weather forced him north and, at 6am on April 19, 1770, a lookout, Zachary Hicks, became **the first westerner to see eastern Australia** when he spotted land (now called 'Point Hicks', Victoria, roughly half way between Melbourne and Sydney.) Cook turned north, mapping the coastline and landing at Botany Bay (near modern Sydney). He named the area 'New South Wales' and it was here that Britons first met Aborigines: the meeting ended with spears and musket fire being exchanged, but no injuries to either side.

Putting back to sea, the voyage almost ended in catastrophe in April 1770, when the *Endeavour* ran aground on the Great Barrier Reef's coral shelves. To refloat her, Cook ordered that the ship's iron and stone ballast, her drinking water and all but four of her guns be tossed overboard. It took two days to get her moving – and then they found themselves in a leaking boat, foundering in shark-infested waters, 24 miles from the shore of a strange land on the far side of the world. Desperately pumping out seawater, the men managed temporarily to seal the hull and they made it to land where they took seven weeks to complete repairs. Again, they interacted with Aborigines, but peacefully, and the word 'kangaroo' entered our language.

Three years after leaving Plymouth, Cook sailed back into Dover on July 12, 1771.

Thirty people had died along the way (though none from scurvy, thanks to Cook's forward-thinking insistence that they gathered fresh fruit whenever they touched land). In his hold, he had **an astonishing array of never-before-seen plants** like bougainvillea, eucalyptus, acacia and mimosa – he and his botanist **Joseph Banks** (see *Kew Gardens* and *Tea*) became the talk of society.

The following year, now promoted to 'Commander', Cook led two ships in a second expedition south in further search of *Terra Australis Incognita*. His son George was born five days before he left, aboard *HMS Resolution* on July 13, 1772 – imagine his wife bidding him farewell, not knowing whether he'd ever come back. He did return, in July 1775, after another three years and 18 days at sea. He had finally proved that there was no unknown continent, and in doing so lost only four men, charted Tonga and Easter Island and discovered many new lands, including the South Sandwich Islands and South Georgia. He also became **the first person to cross the Antarctic Circle** and proved the worth of **John Harrison**'s new chronometer as a navigational tool (see *Longitude Problem*). He brought back a young Ra'iatean man, Omai, who became the first Pacific Islander to visit Europe and was a huge hit in society (and travelled back to the Pacific with Cook on his third and final voyage).

Cook, now a Captain, undertook that final voyage in July 1776. This time, his plan was to discover the fabled Northwest Passage. Again aboard *Resolution*, he found more unknown lands – including Hawaii (which he named the Sandwich Islands) – and mapped great swathes of the American northwest coastline for the first time, though he was unable to find a sea route across. In January 1779, he landed at Kealakekua Bay, one of the Hawaiian islands in the mid-Pacific. The sailors were well-received – some reports say that the natives believed Cook was a God – and they stayed for a month. Upon departure, the *Resolution* encountered a violent storm and was forced to return to the island for repairs. This time, an argument broke out between the Hawaiians and the crew, and one of Cook's small boats was stolen. In an ensuing mêlée, Cook was hit over the head and stabbed to death, and four of his men were also killed. The islanders cut the flesh from Cook's bones and burned it, as was their custom, but some of his bones were later recovered and buried at sea.

Cook was survived by his wife Elizabeth Batts, who bore him six children, three of whom died in infancy. None of the others married, so he has no descendants, but his legacy lies in the map of the Pacific which he left behind, and which has remained virtually unchanged to this day. **He discovered more of the earth's surface than anyone**, and truly succeeded in his aim of going 'farther than any man has been before'.

THE GREAT EXHIBITION

ON New Year's Eve 1999 the Millennium Dome opened in London.

It was a colossal disaster. Tony Blair had said the project – which cost a staggering £789 million – would be 'the most spectacular celebration anywhere in the world'. Instead, it looked like a giant tent with 12 birthday cake candles sticking out of the roof and was described as a 'national laughing stock' by *The Times*. It never turned a profit, and only attracted half the number of visitors originally estimated.

Almost 150 years earlier, our Victorian ancestors had showed the world how it *should* be done.

The Great Exhibition of 1851 attracted more than six million people – a quarter of the Britain's population at the time – in just over five months. Due to advance ticket sales, it was in profit before the doors opened for the first time on May 1. The proceeds, roughly £186,000 (the equivalent of £450 million today, as a percentage of GDP), were used to found lasting monuments including the Victoria and Albert Museum, the Science Museum and the Natural History Museum.

The exhibition was conceived by Prince Albert to demonstrate the ingenuity and resourcefulness of a nation transformed by the Industrial Revolution. The project almost foundered after the initial design for the show hall came in with an estimated cost of £150,000 in materials alone. The landscape gardener **Joseph Paxton** (1803-1865) suggested a new design built almost entirely out of glass. He was given nine days to draw up plans, and came up with a sketch of the 'Crystal Palace' (which can today be seen in the V&A). It was 1,848ft long, 408ft wide and 108ft high, and enclosed 19 acres, and used 293,000 panes of glass, all hand-made in Birmingham, and more than 4,000 tons of iron framework. It was built in just eight months, on time and within budget. (It cost £79,800, or £200 million today.)

Charles Darwin, Charlotte Brontë and George Eliot were among the millions who descended on Hyde Park where Crystal Palace was erected. The magnificent glass exhibition hall contained more than 13,000 exhibits from Britain and her colonies, including the world's first fax machine, huge steam locomotives, clocks, fabrics, false teeth, scientific instruments, kitchen appliances, jewellery, farm tools, pottery, the world's largest diamond and a stuffed elephant. Queen Victoria described the opening as, 'the greatest day in our history, the most beautiful and imposing and touching spectacle ever seen... It was the happiest, proudest day in my life'.

When the exhibition closed in October 1851, the Crystal Palace was moved, piece-by-piece, to Sydenham where it remained as a tourist attraction until it was destroyed by fire in 1936

THE GUILLOTINE

FEW things are more 'French' than the guillotine – but it was actually designed and used in Britain 500 years before it crossed the Channel. Originally called the 'Halifax Gibbet', it's another – albeit rather gruesome – example of British ingenuity at work.

It was invented in Halifax in the 1200s, where the town's 'Gibbet Law' gave the Lord of the Manor the power to execute anyone for thefts to the value of 13½ pence or more. The gibbet was a 15ft high wooden structure with an axe-shaped blade at the top. An executioner would cut the rope holding the blade in place and it would crash down onto the criminal's neck, severing his head. (For the purposes of poetic justice, if the offender was to be executed for stealing livestock, the rope was fastened to a pin holding the blade in place and then tied to the animal in question; it would then be led away, yanking out the pin and causing the blade to plummet.)

Beheadings took place on market days, guaranteeing an audience, and the first recorded use of the Halifax Gibbet was in 1286 when 'John of Dalton' was executed (sadly, no record remains to explain why). That famous British sense of fair play was much in evidence, however, and criminals *were* given a sporting chance. If a crook could escape on the day of execution and make it across the town's boundary line, he was free – as long as he never returned. One John Lacy apparently managed to achieve this, running to freedom in 1616. He returned to the town seven years later, thinking bygones might be bygones, but was recognised and beheaded.

A total of 53 people, including five women, were executed on the Halifax Gibbet between 1541 and 1650 when it was used for the last time. The French started using the guillotine in 1789, the better to remove the heads of the bourgeoisie during the revolution.

HABEAS CORPUS

THE idea that the government cannot simply throw you in jail if it does not like the cut of your jib is burned into the DNA of every Briton – and it's an idea that has travelled around the civilised world, thanks to a document signed in a field in Surrey in 1215AD.

King John (1166-1216) was forced to accept the Magna Carta (Latin for 'Great Charter') at Runnymede by rebellious noblemen, angry at the way he was running England and her overseas military campaigns. Before the Magna Carta, the king was all-powerful – his word was the law, he could do as he pleased and could deal with his subjects as he wished. The Magna Carta made him – theoretically – subject to the law himself and, importantly, introduced the rule of *habeas corpus*. Translating as 'You have the body', it provides ordinary people with a means of legal challenge against unlawful detention by the state.

Sneakily, John placed only his wax seal on the document – rather than signing it – and renounced it as soon as the barons had left. This plunged the country into a brief civil war which ended when he died of dysentery the following year. The charter was reissued, in shorter form with some powers restored to the new king, Henry III; *habeas corpus* stayed in force, however. Its first recorded use was in 1305, and it was codified in statute by the Habeas Corpus Act

1679. Magna Carta itself is seen by many experts as the foundation of many modern constitutions, including that of the USA.

HALLOWEEN

HALLOWEEN is not an American export – in fact, we sent it there. It was invented by the British 2,000 years ago. At that time, the Celts celebrated New Year's Day on November 1. October 31 marked the end of the summer and harvest and the start of the long dark, cold nights of winter. The Celts believed that, on the last day of the year, the spirits of the dead could return to cause death and destruction, and lit huge bonfires to frighten them away. During the festival, which they called Samhain (pronounced 'sow-in'), they dressed up in animal heads and skins and left food outside their homes for hungry spirits. They also used the embers of the bonfires to light fires in their hearths, believing that it would protect them from evil during the winter.

Pope Boniface IV decided in the 7th century that November 1 would be All Saints' Day, also called All Hallows' Day. The night before was All Hallows' Eve, or Hallowe'en. During the 18th century, other Halloween traditions emerged in Britain. English girls would throw apple peelings over their shoulder and hope that they would fall in the shape of the initials of their future husbands, while in Scotland women wrote the names of their suitors on hazelnuts and then threw them in the fire. Whichever nut didn't explode, and instead burnt to ashes, denoted her true love. (If several nuts burned to ashes instead of exploding, that was more complicated.) Anyway, the Americans did not start to celebrate Halloween until the middle of the 19th century when the Brits emigrated, taking their customs with them.

THE HAIRPIN

ERNEST Godward was a great Briton in many small ways. Despite having only three years of schooling, he became a prolific inventor, sportsman, musician, businessman, entrepreneur and adventurer. A sickly child, Godward (1869-1936) was born in Marylebone, the son of a London fireman, and didn't attend school until he was nine. Aged 12, he ran away to sea and got to Asia before a British consul sent him home. Following an apprenticeship at a London steam fire-engine makers, and a short period as a steward on P&O ferries (where he was champion napkin-folder), he emigrated to New Zealand, aged 17. There, Godward worked for a cycle manufacturer but left in 1900 to concentrate on his own inventions, which included the burglar-proof window, a rubber hair curler, a mechanical hedge clipper and an egg-beater. But it was his spiral hairpin (the spiral meant it stayed in place, unlike its rivals) which made his considerable fortune. He sold the rights to the product to an American company in 1901 for £20,000 – worth more than £8 million today.

Godward was also a champion cyclist, runner, swimmer, rower and boxer. And an accomplished singer and musician. And a very good oil painter. Plus, he won the Invercargill to Dunedin motor race in 1909, and became a leading authority on the internal combustion engine.

In 1908 he invented the 'petrol economiser' – the forerunner of the modern carburettor. London's authorities were interested in fitting the economiser to the city's buses, and he came back to Britain in 1913 to promote it, establishing the Godward Carburettor Company at Kingston-upon-Thames. But no-one could quite believe that a product dreamed up by a man with so little formal education could work, so it failed to take off. Undaunted, he moved to New York and sold his economiser to the US Army in 1926. Trials showed that it allowed vehicles to run on kerosene, gasoline oil, fuel oil and even 'bootleg liquor' as well as petrol, and was also more economical and reduced the risk of fire. Three years later, the City of Philadelphia fitted 580 buses and 3,000 taxis with Godward's economiser.

Even on the day that he died, he excelled at something. On December 2, 1936, sailing home to see his family (he was a father of 10), he expired from a heart attack after winning an onboard skipping competition.

HEART, LUNG AND LIVER TRANSPLANT

SURGEONS in Cambridge carried out **the world's first triple organ transplant** in 1986. Davina Thompson, a miner's wife from Rotherham, Yorks, had the operation to give her a new heart, lung and liver at Papworth Hospital. She was 35 at the time, weighed just five stones and was confined to a wheelchair. The procedure, which was carried out by British surgeons **Sir Roy Calne** and **Professor John Wallwork**, was successful and gave her an extra 12 years of life. She died of lung disease at the same hospital in 1998. Her healthy heart then went to another patient. Sir Roy Calne said: 'She was very strong-minded and it was her bravery that prompted us to carry out the operation.'

HIP REPLACEMENTS

HIP replacements were first attempted at the end of 1800s with gold and ivory being used in place of the ball at the top of the thigh bone. But it's thanks to the pioneering work of a Lancashire orthopaedic surgeon, **Sir John Charnley** (1911-1982), that around 50,000 people in the UK – and an estimated 750,000 people worldwide – are relieved of hip pain each year. The pharmacist's son, a bachelor until the age of 46 (he met his wife Jill on a ski slope and married her three months later), was devoted to his patients.

He established a centre for hip surgery at Wrightington Hospital, Wigan, in 1961, where his **groundbreaking total hip replacement** work was carried out using the 'Charnley prosthesis'. At the time, partial hip replacement operations were being performed in Europe and the US with varying success. Surgeons knew little about the body's acceptance of artificial materials, and Charnley experimented with various substances, including Teflon, before settling on a form of high density polythene for his replacement joints.

Knighted in 1977, Charnley also did considerable research into acrylic bone cement which is used to secure the artificial joint to the healthy bone and made huge strides in reducing the potential for infection in his patients, designing sterile-air operating enclosures and using dedicated clean instrument trays to prevent cross-infection. He kept a close eye on each patient and wrote to each of them to ask for the artificial hips back when they died so he could see how they had performed and how well the cement was working. Knee and shoulder replacement surgery developed directly out of his work on the artificial hip, making his an outstanding contribution to the relief of human suffering.

HOCKEY

THE *Encylopaedia Britannica* says hockey was first played by 'ancient civilisations', but London2012.com is clear: 'the modern sport of Field Hockey was developed in the British Isles as an alternative to Football for cricketers seeking a winter sport in the mid-19th century'. It's **the world's oldest stick-and-ball game**, and is India's national sport, having been spread there by the British Army during the days of the Empire.

THE HOVERCRAFT

SIR Christopher Cockerell's father once described him as 'no better than a garage hand' – a slightly unfair way of describing the man who **gave the hovercraft to the world**.

Born in 1910 in Cambridge, Cockerell was an inveterate tinkerer and inventor who had worked on radar during WWII. After 1945, he opened a small boatyard in Norfolk, where he spent his days messing around in his engine shed (much to the disappointment of his distinguished museum director dad).

The boatyard was not a roaring success, and he spent long hours looking at the ships sailing by, just thinking. Ships are not particularly quick; this is because the friction of the water against their hulls holds them back. Cockerell wondered idly whether compressed air, if somehow blown down over the sides of a boat, might not lift it slightly from the water and reduce that drag. From this, he moved to the idea of a vehicle floating on a cushion of air.

The concept of using a fan to blast out high pressure air downwards, allowing a vehicle to float on it, had been thought of before, but Cockerell's breakthrough was to realise that, instead of pointing the jet straight down, the air from the fan should be directed as a jet around the edges of the craft, blowing inwards. That would reduce the amount of air escaping from underneath the vehicle and therefore increase the air pressure lifting it. In classic British fashion, he proved his hypothesis with a working model made from a hairdryer, an empty Kitekat can, an old coffee tin and some rubber tubing.

He patented the concept in 1953 and immediately had his idea restricted by the authorities, who decided it was of possible military value. But in 1958, the National Research and Development Agency bought his design and built a prototype; by 1959, this was crossing the English Channel between Dover and Calais.

Cockerell coined the name 'Hovercraft' and a colleague, Denys Bliss, came up with a flexible skirt around the base of the craft to improve its stability. Initially successful – the record for a cross-Channel route is 22 minutes – they became commercially unviable after the Channel Tunnel opened.

HUMOUR

HUMOUR doesn't always cross national boundaries, and there is no way of judging jokes. We can't *prove* we're the funniest nation in the world. But there is such a thing as the British sense of humour – a dry, sarcastic wit that may actually be genetic. Scientists at the University of Western Ontario studied genetic and environmental contributions to humour in nearly 2,000 pairs of UK twins and a US study examined the humour of 500 sets of North American twins. Dr Rod Martin, one of the Canadian researchers, found surprising differences. 'In North American families, there was a genetic basis to positive humour, but negative humour seems to be entirely learned,' he told *The Independent*.

'In the UK, both positive and negative styles had a genetic basis in the sample. The genetic basis to negative humour in the UK was close to 50 per cent. Certainly in the UK, TV humour is more biting, whereas in North America it tends to be blander.'

This may explain our love of caustic humour, as found in *Fawlty Towers*, *Blackadder* and *The Office*. Ever since the days of *Monty Python*, British comedy like this has been exported to help the humour-free all around the world.

In 2002, **Dr Richard Wiseman** of the University of Hertfordshire attempted the gravely important task of identifying the world's funniest joke. Wiseman and his team of highly-trained scientists asked people around the globe to send in their jokes, and to judge those of others. By the time the experiment finished, the researchers had evaluated over 40,000 gags and almost 2 million votes had been counted.

The funniest joke, unsurprisingly, was sent in by a Briton, **Gurpal Gosall** of Manchester. It goes as follows: A couple of hunters are out in the woods when one of them falls to the ground. His eyes are rolled back in his head and he doesn't seem to be breathing. The other whips out his mobile phone and calls the emergency services. He gasps to the operator: 'My friend is dead! What should I do?' The operator, in a calm soothing voice, says: 'Just take it easy. First, let's make sure he's dead.' There's a brief silence, followed by the sound of a gunshot. Then the man comes back on the line. 'OK,' he says. 'Now what?'

The funniest joke according to *UK-only* votes was this: A woman gets on a bus with her baby. The bus driver says: 'That's the ugliest baby that I've ever seen!' The woman goes to the rear of the bus and sits down, fuming. She says to a man next to her: 'The driver just insulted me!' The man says: 'You go back up there and tell him off, love – and don't worry, I'll hold your monkey for you.'

Tee hee.

HYDROELECTRICITY

AN INGENIOUS British entrepreneur's home was **the first in the world to be powered by hydro-electricity**.

William Armstrong, who lived with his wife Margaret in Cragside House, Rothbury, Northumberland, installed the supply in 1878. It powered labour-saving devices including an early dishwasher, a roasting spit, an hydraulic servants' lift and even a plant-pot-turning device to ensure that the peaches in his greenhouse ripened evenly. When by Christmas 1880 Armstrong's house was lit with Joseph Swan's new-fangled incandescent light bulb, it became the second house in the world to be lit by electricity – Swan's was the first.

Described as 'The Palace of the Modern Magician', Cragside – which also had central heating and running hot water – was built into rocks on a rugged hill. Man-made lakes in the grounds were dammed and fed down to provide the flow needed to power a dynamo. Armstrong, who was born in Newcastle in 1810 and planted seven million trees and bushes on his estate, was way ahead of his time. In 1863, he forecast that solar and water power would become increasingly important and predicted that 'England will cease to be a coal producing country... within 200 years'. He gave up work as a lawyer after inventing a hydraulic crane which cut the turnaround time of ships at Tyneside quays, increasing revenue for the local council, and later moved into the arms and ship-building business, inventing a lightweight long-range gun after hearing that British troops in the Crimean War were being hampered by the heavy weapons in the field. He died a millionaire at Cragside House, aged 90. It is now owned by the National Trust and open to visitors.

HYDROGEN

HYDROGEN is absolutely essential to life – it is the 'H' in H_2O (water), and it comprises around three-quarters of the mass of the sun (in fact, it forms 75% of the mass of the universe).

It was identified as an element by a strange, shy British scientist called **Henry Cavendish** in 1776.

Others had produced hydrogen before (you can do this by mixing a metal such as zinc with a strong acid) without realising what it was. Cavendish isolated the gas and called it 'inflammable air'. He also observed that it reacted with oxygen to form water and later accurately determined the composition of the earth's atmosphere (roughly 80% was 'phlogisticated air', which we now know as nitrogen and argon, and around 20% was 'dephlogisticated air', now 'oxygen').

Cavendish was born into an aristocratic British family in Nice, France, in 1731, and educated in Hackney. A quiet and introverted man who was apparently terrified of women, he had a back staircase installed at his house so that he could sneak out in the unlikely event that any should pay him a visit. After his death in 1810 many personal papers were found in which he suggested theories which others had come up with much later than he. For instance, Cavendish discovered that electrical current is proportional to voltage almost 50 years before the German Georg Ohm came up with what is today known as Ohm's Law.

HYPNOSIS

WHEN the Scottish surgeon **James Braid** watched a travelling mesmerist's act in Manchester in November 1841, he dismissed him as a charlatan. 'I saw nothing to diminish my previous prejudices,' he said.

Strange, then, that the doctor born in Fife in 1795 is now considered the father of hypnosis. Within a year of denouncing the showman, Braid was a convert. He had experimented by putting his wife, friends and servants into trances, and had realised that there was a scientific basis for the phenomenon after all. He coined the name 'hypnosis' after Hypnos, the Greek god of sleep, and published a book with the catchy title *Neurypnology, or The Rationale of Nervous Sleep Considered In Relation With Animal Magnetism*, speculating as to the possible therapeutic affects of hypnosis, particularly whether it could be used to alleviate the pain and anxiety of patients in surgery. Within three years, James Esdale, a doctor with the East India Company, was indeed using hypnosis in surgery, and carried out many successful operations without the aid of anaesthetic. By 1847, Braid had discovered that his subjects were not actually asleep during hypnosis and tried to rename it monoideism, meaning a state of prolonged absorption in a single idea, but by then the words 'hypnosis' and 'hypnotism' were in common usage.

THE HYPODERMIC NEEDLE

UNTIL Edinburgh doctor **Alexander Wood** (1817-1884) invented the hypodermic needle in 1853, drugs had to be administered orally or through a cut in the skin.

The son of a doctor, Wood was experimenting with needles, including those used for acupuncture, when he made a hollow one thin and sharp enough to pierce the skin. **The first person to receive an injection** was given morphine by Wood to treat neuralgia: he noted that the patient fell into a deep sleep. He started to use his needle to inject himself and his wife, Rebecca Massey, with morphine. Tragically, she became addicted, and his invention led to her becoming the first person to die of an intravenous drug overdose.

In 1946, the Birmingham firm Chance Brothers invented **the first all-glass syringe** with interchangeable barrel and plunger which allowed mass-sterilisation of components (without the need to match an individual barrel to its plunger).

(The French also claim credit for the invention of the hypodermic needle – Charles Pravaz was independently pioneering his version at the same time as Wood.)

IBUPROFEN

A BRITISH railwayman's son who left school at 16 to work at Boots chemists went on to create one of **the world's best-selling**

painkillers. Working in the front room of a Victorian house in Nottingham in the 1950s, **Stewart Adams** came up with ibuprofen. He created it as an anti-inflammatory painkiller for people with rheumatoid arthritis – but he soon realised its further potential after celebrating his success with one too many drinks at a conference. He said later, 'I had a hangover, so I took 600mg of ibuprofen.' The results were 'magic', he said. The drug was patented by Boots in 1962 and it was available as a prescription-only medicine seven years later.

It had taken 16 years for Dr Adams and his colleague, John Nicholson, to come up with a drug that worked – their earlier efforts had no potency, produced rashes or induced jaundice. Ibuprofen became available to buy over the counter in the early 1980s. By 1985, it was estimated that more than 100 million people had been treated with the drug in more than 120 countries. Now in his 80s, Dr Adams was given an OBE for services to science in 1987.

THE INDUSTRIAL REVOLUTION

MUCH of what made Britain 'great' would never have happened without the Industrial Revolution – a new age of technology and mechanisation where we led the world. This period, roughly speaking from 1700 to 1900, was underpinned by four things: coal, steam, iron and steel (though textiles were also important). It was a time of wild eccentrics like 'Iron Mad' John Wilkinson and cool-headed visionaries like Isambard Kingdom Brunel – when unimaginable fortunes were forged and lost in hissing, clanking, smoking factories such as the world had never seen before.

Coal and Steam

IN 1700, our largest pits were 300ft deep – the height of St Paul's Cathedral – and produced 2.5 million tons of coal a year. This was nowhere near enough to fuel the Industrial Revolution – by 1900, we would require *250 million* tons a year. The only way to produce huge quantities like these was from the thicker, richer seams much further underground. By the late 1800s, pits were 2,500ft deep. But this made the miner's job more dangerous – flooding and the build-up of explosive and poisonous gases worsened the further down they went. Previously, teams of horses – sometimes hundreds strong – hauled up floodwater in buckets on ropes. But this was slow and expensive, and an industrial revolution could not be driven by horses. Mine owners were desperate for an alternative.

In 1698, the Devon military engineer **Captain Thomas Savery** provided the beginnings of the answer, when he invented **the world's first practical steam engine**. Called the 'Miner's Friend', it heated water in a boiler and collected it in a container. As the steam cooled it condensed, shrinking in volume and leaving behind a partial vacuum which sucked water up through a pipe. However, even a perfect vacuum can only raise water a short distance, and the worst-affected mines were the deepest. Additionally, the Miner's Friend was dangerous to use. The French scientist John Desaguliers reported perhaps the first death of the early industrial age when an inexperienced worker overpressurised an engine and 'after some time, the steam… burst the boiler with a great explosion, and killed the poor man'.

Clearly, the laws of physics precluded a practical 'sucking' engine. What was needed was someone to invent a device that *pumped* rather than pulled the water upwards.

Thomas Newcomen was an ironmonger and radical protestant preacher. Born in 1664 in Dartmouth, Devon, he also lived only 15 miles from Thomas Savery, and had helped him build the first Miner's Friend. In return, he kept one for himself, and he spent time tinkering with and improving it. He knew that flooding was a problem in local tin mines and, in 1712, he launched his 'Beam Engine' on the world. As with Savery's model, it started with hot steam condensing and shrinking to form a partial vacuum, but then it used atmospheric pressure to drive a piston connected to a pivoting beam which, in turn, raised pumping gear at the bottom of a mineshaft. Although it was inefficient and developed only around 5.5 horsepower, it was better and cheaper than any previously available method of draining pits. A W Skempton's *Biographical Dictionary of Civil Engineers in Great Britain and Ireland* records that each stroke lifted 10 gallons of water from a depth of 153 feet, or just over 46 metres. Importantly, the engines were strongly-built, reliable and easy to repair if they did go wrong and they could be operated 24 hours a day.

By the time Newcomen died in London in August 1729, aged 65, there were at least 100 of his engines working in Britain and Europe. (They were still used into the 20th century, and one was still operating 127 years after it was installed.) Once mines could be drained to greater depths, more coal and metal ores could be extracted more quickly: this was vital to the expansion of industry and the Beam Engine was, thus, one of the most important drivers of the Industrial Revolution.

Despite this, Newcomen received little cash or credit for his invention, with most of the limelight shining on **James Watt** (1736-1819). Watt, a prematurely grey Scots inventor with a dry wit and a self-deprecating style, earned his place in history when he worked out how Newcomen's engine could be vastly improved, and used for more than merely pumping out flooded mineshafts.

When he was born, in Greenock on the Firth of Clyde, Newcomen's engines were already wheezing and pumping throughout the land. Watt, considered weak and 'slow' as a boy, had excelled at mathematics and started work as an instrument maker at Glasgow University in 1758. He developed an interest in steam power, and constructed prototype engines of his own with little success. In 1763, though, he learned that the University owned a Newcomen engine which needed repairing. He asked if he could carry out the repairs and dismantled it for study, identifying the flaw in the Newcomen design which prevented it from being particularly efficient.

In Newcomen's engine, the vacuum inside the cylinder required to drive the process was formed by condensing the steam using a water spray inside the cylinder. The cylinder was then re-heated with fresh steam and the process repeated. Watt realised that the repetition of this hot-cold-hot process on each cycle was a huge waste of energy. If you diverted the spent steam out of the cylinder and into a separate 'condenser' chamber for cooling, the cylinder itself would stay hot and be far more efficient. Additionally, he saw that the steam in Newcomen's engine was not being 'worked' particularly hard. Newcomen's cylinder was effectively open at the top, sealed only by the piston. Thus the steam was at little more than atmospheric pressure. Watt saw that if you sealed the cylinder and increased the pressure inside – albeit that the engine remained a low pressure vessel – and introduced steam at the top and bottom alternately, you would make Newcomen's design far more efficient.

He patented his improved engine in 1769 and formed a partnership with **Matthew Boulton**, a Birmingham entrepreneur who had ironworkers able to precision engineer large cylinders and tight-fitting pistons (virtually unknown at the time).

As well as using his engines for pumping, Watt also adapted them for grinding, weaving and milling; this was **the world's first serious mechanisation** of processes previously done by hand, and it vastly increased their speed and efficiency.

He was renowned as a serious man, cautious, plagued by self-doubt despite his brilliance, and modest – he declined a baronetcy and the high sheriffships of Staffordshire and Radnor (where he had homes). He made many other breakthroughs – the 'rev counter' and the unit of horsepower are among those still in use today – and retired in 1800, having made a fortune (though his life was scarred by tragedy; his first wife died in childbirth, leaving him with two young children, and his second marriage produced a son and a daughter who both died young of 'consumption' – tuberculosis). He died in Handsworth, Birmingham, in August 1819, and was buried beside Boulton. The unit of electrical power was named the 'Watt' in his honour.

Although Watt's designs were groundbreaking, they were cumbersome and relied on low pressure, and thus could not produce really serious amounts of power. **The world's first practical high-pressure steam engines** – regarded as dangerous by Watt – were developed by Cornishman **Richard Trevithick** (1771-1833).

Nowadays, boys like Trevithick are given ASBOs or force-fed Ritalin – his school reports describe an obstinate, spoiled and disobedient child with a reputation for truancy. But he was a quick-witted lad with a good head for figures. His father was a copper mine manager, and the young Richard would have been familiar with his battles to keep floodwater at bay. He was born as Newcomen's engines were being replaced by Watt's improved version, and throughout his childhood Cornish miners were striving to find ever better ways of pumping out their pits (and to get round Watt and Boulton's patents).

Trevithick grew into a giant of a man, a 6ft 2in wrestling champion who was said to be able to hurl a sledgehammer over a roof, and started work in a local mine aged 19. He spent the 1790s pondering the limitations of Watt's low pressure systems. By 1800, while working as an engineer at the 'Ding Dong' mine near Penzance, he had invented a high pressure steam engine.

His design used the steam itself to move his piston, instead of relying on atmospheric pressure to do the job as 'condensing' engines did. Trevithick's engine was expensive to construct, but it did not need a condenser and it used less coal. It also produced what were, for the time, phenomenal amounts of power. A small 1802 model operated at a staggering 145 pounds per square inch – by comparison, modern car tyres are pressured to about 30psi – although the downside of this was an increase in danger. In 1803, an engine being used to pump water from the foundations of a building in London exploded, killing four men. The tragedy was actually caused by human error, but 'strong' steam's detractors – including Watt and Boulton – were quick to denounce its use as a perilous folly. Undaunted, Trevithick introduced numerous additional safety features, some of which are still in use on steam engines now.

His new engines were known as 'puffers' – condensing engines were relatively quiet – and by 1812 he had blended his designs and Watt's to create the prototype 'Cornish Engine'. This used high-pressure steam in both expanding and condensing modes, and was eventually known and used around the world.

The problem of flooding in mines was effectively solved, and coal could be hewn from ever deeper seams. But the deeper a mine went, the greater the build up of heavier-than-air gases like methane and carbon monoxide. Methane, which occurs naturally in coal seams, is highly explosive, particularly when mixed with coal dust. The flickering flame from a miner's candle could set off this mixture – called 'fire damp' – with terrible effects. For instance, the Tyneside Felling Colliery explosion of 1812 killed 91 men and boys, many of them fathers and sons working side by side. Equally deadly, colourless and odourless carbon monoxide, created by underground fires or explosions, could suffocate miners.

Cornish-born **Humphrey Davy**'s 1815 invention of the **safety lamp**, which enclosed a naked flame inside an iron gauze, reduced the risk from fire damp explosions. Canaries were used to alert miners to the presence of carbon monoxide. The birds, which are very sensitive to that gas, were kept in cages underground; at the first sign of a canary becoming distressed, a pit would be evacuated (it wasn't until 1986 that these birds were replaced by electronic gauges in British mines). Meanwhile, steam-powered fans were also introduced to ventilate shafts, greatly reducing the dangers from these gases. Mining itself became mechanised, too. Pick axes were supplemented by steam-powered tools and the coal was hauled to the surface in ever greater quantities.

Steam Cars

STEAM could be used for much more than pumping out mines. Factories were no longer dependent on rivers and water mills for their power so could be sited almost anywhere – leading to the growth of towns and cities. But it could also be used for generating motion.

Some time around 4,500BC, man had tamed horses and developed the wheel. Perhaps two thousand years later, someone had put the two together and the horse-drawn cart had been born. But that was pretty much as far as things had gone: the haywain painted by John Constable in 1821 would have been recognised for what it was by the Romans who had roamed Britain two millennia before. But not long before Constable mixed his pigments and oil and sat down to paint in Suffolk, a revolution had taken place two hundred miles away in Cornwall. It was every bit as dramatic as the discovery of the horse and the wheel thousands of years earlier.

On Christmas Eve in 1801, Richard Trevithick demonstrated **the world's first practical automobile** (the Frenchman Nicolas-Joseph Cugnot had earlier built a steam tricycle but it had a top speed of 2.5mph and was dangerously unstable). Trevithick called his steam-powered road carriage 'The Puffing Devil', and used it to carry several (brave) men up a hill in Camborne and beyond to the village of Beacon, a mile or so away. We can only guess at the excitement those early pioneers of horseless transport must have felt, albeit travelling at only a few miles per hour, and the consternation of the local people as this clanking, gasping, shuddering contraption hove into view. A few days later, the Devil was driven the two or three miles to Tehidy, just inland from the north Cornwall coast. Unfortunately, it broke down on the rough roads *en route* and when its operators adjourned to the pub for a festive drink and left the fire burning, the whole thing melted.

In 1803, Trevithick built a second locomotive, which he called 'The London Steam Carriage'. To show it off, he took a group of passengers on a trip through the capital between Paddington and Islington, with the streets specially closed to horse-drawn traffic. It caused widespread admiration, though the passengers complained that it was uncomfortable – despite the eight-foot wheels designed to smooth the journey – and the following day he crashed it into some house railings in Britain's first automobile accident. This and the fact that it was expensive to run – it required two men and a hundredweight of coal per mile, whereas a Hackney carriage needed only one man and a bale of hay – killed it as a commercial project. But Trevithick had undoubtedly shone a light into the future.

Steam Trains

PUTTING road travel to one side, Trevithick wondered whether steam could be used to power a vehicle along rails (early railways featured horse-drawn wagons on wooden tracks). On February 21, 1804, he demonstrated **the world's first railway locomotive** at Penydarren ironworks near Merthyr Tydfil. It hauled a load of 10 tons of iron, 70 men and five wagons for a distance of 9½ miles, at an average speed of 5mph. Although it broke the rails and was not used again, it is impossible to overstate the scale of this breakthrough: it spelled nothing less than the end of the era of the horse, and a new age in industrial transport.

In 1808, Trevithick ran a new passenger locomotive, the 'Catch-me-who-can', on a circular railway in London. Thrilling 12mph rides were available at one shilling each, but – again – broken rails meant it was not a financial success. (This chicken-and-egg problem dogged early railwaymen: there were no trains, so the available rails were not designed for trains, so trains would not run on them.)

In 1817, he went to Peru to advise on the use of his engines in the silver mines there. He ended up owning several mines, but the war of independence with Spain broke out across much of South America and ruined everything. He had to flee the country, leaving behind £5,000 worth of ore (worth £3.5 million today). On his perilous journey home he almost drowned crossing a river, narrowly escaped from an alligator and nearly starved. After a decade of adventures in South America, the *Oxford Dictionary of National Biography* records, he arrived back with nothing more that the clothes he stood up in, a gold watch, two compasses and a pair of silver spurs.

His luck did not improve in England. He tried and failed to claim money from others who had abused his steam engine patents during his time abroad, and failed also with further ingenious inventions, including one for an early storage heater and another for a ship propelled by water-jet, before dying, penniless, in 1833. In this, he was like so many great inventors who, through lack of business acumen, sadly did not manage to capitalise on their genius.

In the meantime, steam-powered travel moved relentlessly on. **The world's first railway** had been set up in Leeds as far back as 1754, using horses to haul coal from the nearby Middleton Pit to the centre of the city. In around 1807, iron rails replaced the wooden ones and in 1812 it became **the world's first *steam* railway** when it switched to a locomotive based on Trevithick's Catch-me-who-can. By 1830, the world's first inter city passenger railway line – the Liverpool and Manchester Railway (L&MR) – was opened, with Northumberland-born **George Stephenson**'s famous 'Rocket' winning a competition to select the engine to be used on the route. Rocket could cover the 50-mile round trip at 12mph while pulling 13 tons, and 29mph while running light. The opening day of the L&MR, on September 15, also saw the death of the Liverpool MP William Huskisson, who was knocked down and killed by the locomotive, but despite this tragedy it was the start of a rail network that soon linked every town in Britain to the docks and factories which – along with our great canals – fed the industrial age.

Steam Ships

STEAM also powered our military and merchant shipping for the next century. There were many great British shipbuilders in the steam age.

Isambard Kingdom Brunel was an engineering polymath born to a French father – the almost equally brilliant Sir Marc Brunel – in Portsmouth in 1806. By the age of 20, Isambard was helping his father tunnel under the Thames (Sir Marc finished that project in 1842), and in the 1830s he designed the first of many bridges (he was responsible for the 702ft Clifton Suspension Bridge at Bristol, though died before it was completed). In 1833, he was appointed chief engineer of the new Great Western Railway – a fabulous project, and perhaps his greatest accomplishment – which linked London to Bristol and Exeter. It featured a series of stunning architectural and engineering achievements, among them magnificent stations (including Paddington, intended to outshine Euston, the London home of GWR's rivals, the Great Northern Railway), huge viaducts and the then **longest tunnel in the world** (the nearly two miles-long Box Tunnel, linking Bath and Chippenham). In the Crimean War of the 1850s, he designed **the world's first prefab hospital** for the Crimea. They were hygienic, properly ventilated, had drainage and sanitation and saved the lives of hundreds of injured soldiers.

As well as all of this and more, Brunel built **the world's first modern ships**. Determined to bring in a new era of transatlantic travel, he had designed and launched the *SS Great Western* in 1837. She was a 236ft-long wooden affair with steam powered paddle-wheels and sails, and she made the return trip to New York City in 29 days, compared with two months for a sailing ship. This in itself was a visionary triumph: at the time, no-one believed that a practical steam ship could cross the Atlantic, since it was thought that the fuel required would take up all available space.

But it was his *SS Great Britain* which set the standard by which all other vessels of the period would be judged. She was the **world's first iron-hulled, propeller-driven ship**. At 322ft in length, with her riveted wrought-iron construction contributing to a weight of around 3,000 tons, she was also the biggest thing on the seas when launched in 1843. She had cost £117,295 (£325 million in today's GDP terms) and went on to have a long career, initially taking passengers to America and later to Australia.

Brunel's third major ocean-going liner, the *SS Great Eastern*, was a financial disaster, though mostly this was because she existed at the very cutting edge of technology, and her cost made her doomed to unviability. Almost 700 ft long, weighing 18,915 tons and with **the**

world's first double hull, she was the largest ship in existence until the turn of the century. Capable of carrying over 4,000 passengers all the way around the planet without refuelling, her enormous engines produced 8,000 horsepower – showing how far steam had come since Newcomen's 5.5hp Beam Engine. But the project fell way behind schedule – and soared far over budget – as Brunel and his colleagues struggled with a series of technical difficulties.

Work began in 1854, and her maiden voyage was not until September 1859. By this time, Brunel was seriously ill with kidney disease and he collapsed of a heart attack on the deck of his 'leviathan' two days before she launched. He died ten days later, shortly after hearing of a colossal explosion in her engine room as she steamed along the English Channel. Five stokers were killed and many others injured. An eyewitness account in *The Times* paints a gruesome picture: 'A man blown up by gunpowder is a mere figure of raw flesh, which seldom moves after the explosion. Not so with men blown up by steam, who for a few minutes are able to walk about, apparently almost unhurt, though in fact mortally injured beyond all hope of recovery... This was so with one or two, who, as they emerged from below, walked aft with that expression in their faces only resembling intense astonishment, and a certain faltering of the gait and movements like one that walks in his sleep. Where not grimed by the smoke or ashes, the peculiar bright, soft whiteness of the face, hands or breast, told at once that the skin, though unbroken, had been boiled by the steam.'

The *Great Eastern* survived, though she only made nine transatlantic voyages and never turned a profit – she was simply too big, with not enough demand for the 4,000 passengers she was built to carry, or to fill her cavernous cargo holds. But she proved beyond doubt that giant ships *could* be built, and laid the foundations for all the shipping that followed.

(She also performed one extremely important task, after being converted to a cable-laying vessel in 1865 – see *Transatlantic Cable*.)

Sir Charles Parsons was born a nobleman – the sixth son of William Parsons, third Earl of Rosse – in London in 1854. He was a scientist and inventor who was fascinated by mechanical things from an early age – as a youngster, he and his brothers built their own home-made steam car (tragically, their cousin Lady Bangor fell from it and died while playing with them one day). After Cambridge, Parsons undertook a four-year apprenticeship on Tyneside and then moved to Leeds, where he developed a rocket-powered torpedo and wooed his wife Katherine with, of all things, his skill at needlework.

In 1884, while a junior partner in a Gateshead engineering firm, he developed **the world's first steam turbine**. Steam engines still drove pistons as they had since the days of Newcomen and Watt. But in order to develop ever greater amounts of power, they had become larger and noisier, and, apart from being a public nuisance, they were reaching the limits of the science – eventually, 'bigger' no longer means 'better'. Parsons' turbine worked by extracting the thermal energy from pressurised steam and turning it directly into rotary motion. It was far more efficient, and had a much greater power-to-weight ratio, than the older engines.

Initially, he planned to use it for creating electricity – and, indeed, about 80% of the world's electricity is today generated by steam turbine – but he soon realised that it might also be used to drive ships. Starting with a 2ft model boat on a pond, he worked up to the launch of **the world's first steam turbine-powered ship**, the 44 ton, 100ft-long *Turbinia*, in 1894. You might imagine that the Royal Navy would have been eager to embrace this new technology, but only if you have never acquainted yourself with British governmental bureaucracy. In order to convince them of his design, he was forced to gatecrash Queen Victoria's 1897 Spithead Review (a traditional review by the monarch of the Navy's finest and most modern ships, held off the Isle of Wight). Immediately after the Royal inspection, the *Turbinia* roared unannounced into view, weaving in and out of the warships at an unprecedented speed of 30 knots. In fact, her top speed was *34* knots, while the fastest destroyers of the day could manage only 27. Within a few years, both British naval warships and civilian vessels were powered by steam turbines. Among the latter was Cunard's famous *Mauretania*, which was the fastest ship across the Atlantic for a quarter of a century. 'It is said that Charles Parsons sketched the original design for the reaction blades in his turbine on the back of an envelope,' says Cambridge University's Department of Engineering website, 'and this remained the standard for many years. It took £100,000 of research to improve its efficiency by 2%, a testimony to the instinctive knowledge of a true genius... one of the greatest engineers that this country has ever produced.'

Away from the sea, he was a shy, retiring family man, who built steam-powered model helicopters and aircraft as toys for his children. His tireless mind also turned itself to the manufacture of high quality optical glass for telescopes and other scientific instruments, to the 'auxetophone' (an early loudspeaker for increasing the sound of stringed instruments, particularly of the double bass, which was not taken up by sniffy classical musicians) and to a 25-year search for a way of making artificial diamonds. This latter quest is the only one in which he ever failed.

Iron and Steel

NOW, back to the early days of the Industrial Revolution. This age of vast upheaval and progress needed more than just coal and steam – it also needed iron and steel, because without these, you cannot construct steam trains, enormous ships and towering factories.

And consider this: in 1700, Britain produced just 25,000 tons of iron. By 1800, that had risen to 250,000 tons. By 1900, it would stand at an estimated *ten million* tonnes. So the importance of iron – and of those who invented the processes for making and refining it in sufficient quantities – is obvious.

We're actually talking about three kinds of iron.

The first, **wrought iron**, is a relatively soft and malleable substance. It had been used in Britain since around 800BC, and was originally made on a small scale by heating iron ore in charcoal fires. Heating drove off impurities from the ore, and those that were left behind – the 'slag' – were hammered from the metal as it was worked (or 'wrought'). Importantly, the iron absorbed only a small amount of carbon in the heating process – less than 1% – and that low carbon content gave it the malleability which made it perfect for turning into nails, tools and weapons.

The second, **cast iron**, did not arrive in Britain until the middle ages. Called 'cast' iron because it melted and could be poured into pre-prepared moulds for cooking pots, say, or cannon balls, it was much harder than wrought iron – but also very brittle. This meant it couldn't be turned into nails, for instance, because they would shatter when struck. But it was excellent for things which needed to withstand great pressures – like the cylinders of Watt's steam engines, or the struts and supports of Brunel's great bridges.

Its manufacture was more 'industrial' in nature: it was made in furnaces at much higher temperatures, and that extra heat made it absorb far more carbon – around 4% – giving it its hardness. It could be stored as ingots of 'pig iron', so-named because the liquid metal was poured into large sand troughs with smaller troughs attached, and these resembled a sow suckling a litter of piglets. These pig iron ingots could be remelted as and when required.

The third kind of iron – **steel** – is strong *and* relatively malleable, which puts it somewhere between the two and makes it ideal for many industrial purposes. It arrived later.

As the Industrial Revolution gathered pace in the 1700s, the early demand was for more cast iron. Blast furnaces increased in size and waterwheel-powered bellows were replaced by new ones blown by Watt's steam engines. The first major breakthrough in cast iron manufacture was made by **Abraham Darby**, a Quaker from Wrens Nest near Dudley, when he built **the world's first coke-powered blast furnace** some time between 1709 and 1711. Coke – coal which has been baked to remove impurities – was cheaper to use and more efficient than charcoal (and its use also slowed the worrying rate of deforestation). Darby's furnace, at Coalbrookdale in Shropshire, turned out seven tons of iron a week – a significant amount for its day. Sadly, he died young, aged just 39, but he left an astonishing legacy. In 1779, his grandson **Abraham Darby III** built **the world's first iron bridge** across a 100ft gorge in the River Severn at Ironbridge, near Telford. In the interim, the whole area was ablaze with coal, steam and ironmongering technology, and many people regard Ironbridge as the birthplace of the Industrial Revolution. A UNESCO World Heritage Site, it is known as 'The Valley That Changed The World'.

Of course, the demand for nails and tools was increasing, too. If a way could be found to turn pig iron into industrial quantities of *wrought* iron, that would be a significant advance. It was achieved by **Henry Cort**, a Lancastrian born in 1740, who developed the 'puddle furnace'. Molten pig iron was stirred by a highly skilled 'puddler', exposing the metal to the heat and gases in such a way that the high levels of carbon found in pig iron were oxidised off. Without getting too complicated, alloys usually have lower melting points than the individual elements they contain. Thus as the carbon content in the cast iron decreased with oxidisation, semi-solid chunks of purer wrought iron began to appear in the molten mass. The puddler gathered these, worked them under a forge hammer and then ran the still-hot wrought iron through new rolling mills. It was a process which produced flat iron sheets of a uniformity and at such speed that it transformed industry.

When the Royal Navy announced in April 1789 that only iron made according to Cort's methods would be allowed in the construction of its ships, this ought to have ensured his future as a very wealthy man. But, alas, he proved once again that being highly inventive does not necessarily translate into riches. Some bad business dealings rendered him bankrupt and Henry Cort died a poor man in 1800, leaving behind his second wife, Mary, and 13 children; it was a fate he scarcely deserved.

Many of the ironmen were eccentrics, but few compared with **'Iron Mad' John Wilkinson**, a man literally *obsessed* with the stuff. His most significant achievement was the development of a new method of making stronger cylinders for the ever-more-powerful steam engines. It sprang in turn from a new way of making cannon barrels. For many years, barrels had been cast ready for use, in one piece with a hole through the centre. But this method often led to 'honeycombing' in the metal which made the cannon somewhat dangerous to use: they had a tendency to blow up while firing, and often did more damage to their own side than to the enemy. Wilkinson (1728-1808) cast his barrels solid – preventing honeycombing – and bored the hole out afterwards. By 1775, he had perfected this art and was producing magnificent cannon which were not only safer but far more accurate than any previously made. But even in a martial nation like Great Britain, demand was limited – especially now that they were near-indestructible. Luckily, steam engineers needed to find ways of creating iron cylinders which could withstand significant pressurisation; Wilkinson, from Clifton in Cumbria, could make them with his cannon-boring equipment.

He also went on to become a major shareholder in the Ironbridge project. The plan *had* been to construct the bridge only partially of iron, but he badgered and cajoled the other shareholders until they agreed to have it made entirely from the metal. He also built the first iron-hulled barge, installed an iron pulpit and windows in his local Methodist chapel and began a collection of iron coffins. A tall, powerful man with a deep voice and a face scarred by smallpox, he slept with many of his servant girls and took up a young mistress at the age of 72, fathering three children by her before dying in 1808. He was buried in one of his iron coffins, with the spot marked by an iron obelisk. He had vowed to return seven years after his death to see that his furnaces were still afire, and many of his former workers turned up in case he did. He didn't.

Many others made advances of varying degrees to enable the industrial revolution to progress. Among them was **James**

Beaumont Neilson, a 'quiet, reflective, unassuming, and earnest man' who was 'kind-hearted and fond of harmless mirth', and left his school in the Gorbals area of Glasgow before the age of 14. He went on to become wealthy thanks to his 'hot blast' smelting process, introduced in 1828. Previously, cold air had been used to draw iron furnaces; Neilson made the seemingly facile discovery that pre-heated air worked better. It reduced the amount of fuel needed (and allowed for raw coal to be used as well as coke), dramatically increasing the amount of iron a factory could produce.

In the same year, the London architect **Henry Robinson Palmer** invented corrugated iron. Made from the newly-available wrought iron sheets, this provided a light, strong and easily-transported way of making prefabricated buildings. Within a decade, it was in use around the world.

But none of these innovators was as important as **Henry Bessemer**, who came up with another faster, cheaper and better British world first – **a new way to make steel**.

Another alloy of iron and carbon – stronger than wrought iron but not as brittle as cast – steel was originally made by a difficult, slow and expensive process called 'cementation'. Bars of wrought iron were layered in stone boxes full of powdered charcoal which were heated for several days, to force the bars to absorb carbon. To make the distribution of carbon even, the bars were then smashed up, relayered into more charcoal powder, and reheated. This produced so-called 'blister steel'. In the 1740s, a Doncaster clockmaker called Ben Huntsman refined the process somewhat by heating the steel in a clay crucible, but steel remained costly and available only in smallish quantities. This was a major problem; the rapid expansion of the railways had created a huge market for steel rails – wrought iron track was always breaking.

Bessemer provided the answer. Born in Hitchin, Hertfordshire, in 1813, he had made a fortune by finding a way to make gold paint out of bronze powder. The money he made from this allowed him to experiment in heavier industry. His 'Bessemer Converter' was patented in 1856. He built a pear-shaped chamber which was tilted so that molten pig iron could be poured in through the top. It was then swung back to the vertical position and a blast of air was blown in through the base. This was done to burn off most of the pig iron's high carbon content and other impurities, leaving behind a mass of purified iron with a low carbon content, otherwise known as 'mild steel'. It was a highly risky process – the Sheffield Industrial Museums Trust, which owns one of only three Bessemer converters left in the world, describes how 'spectacular but dangerous flames and fountains shot out of the top of the converter' once it was fired up. 'Perhaps only an "amateur" such as Bessemer, who had no training as a metallurgist but nevertheless had the "splendid audacity of ignorance" could have devised it,' says the *Oxford Dictionary of National Biography*. But it was also extremely effective; Bessemer could make seven tons of steel in *half an hour*, a huge advance.

As always, there were problems. One major shortcoming was that Bessemer's method did not remove phosphorus – an element found in most iron ores, and one which also left the steel more brittle than was ideal. Iron ore without phosphorus was only found in a few places worldwide – Wales, Sweden and America's Michigan State – and that rarity, coupled with its quality, made it expensive.

In 1876, **Sidney Gilchrist Thomas** discovered that adding limestone to a Bessemer Converter drew the phosphorus out of the pig iron, resulting in phosphorus-free steel. Thomas, a Londoner born in 1850 of Welsh parents, is typical of his age. He was a police court clerk who studied chemistry after work. A teacher told him: 'The man who eliminates phosphorus by means of the Bessemer converter will make his fortune.' So he did. The idea that a court clerk might make a metallurgical advance of world importance by studying in his spare time seems unlikely – if not impossible – today. Incidentally, his refinement also yielded large quantities of phosphorus-rich slag which could be ground down and used as agricultural fertiliser. He was another who died early – aged only 35, from emphysema – but his discovery meant that iron ore from all over the world could be used to make high quality steel.

At around the same time, another Briton, Charles Siemens (who was born in Germany but lived and worked here, and took out British citizenship), developed a rival steel-making process, the 'regenerative furnace'. Slower than Bessemer's method, it produced steel of very high quality, and further added to the dynamism of the industry. It all led to the skyrocketing production of cheap steel across Europe and in America.

In Britain alone, the figures are astonishing.

In 1855, when steel was still being made using Huntsman's 'crucible' method, Britain's total output was a mere 50,000 tons.

By 1870, British companies were manufacturing over 300,000 tons a year.

By 1900, production was at five million tons.

Siemens and Bessemer were both knighted, and both made fortunes. When Siemens (who founded the international engineering giant which bears his name) died from heart disease and pneumonia in 1883, he left an estate worth £30 million today (on a retail price index comparison) and £200 million if compared with average earnings.

Bessemer was even more successful. He was not universally loved – some suggest he was a raging egotist – and he not everything he turned his hand to worked. For instance, his hugely-complex design for an anti-seasickness platform on ships, involving suspended seating, hydraulics and a spirit level, flopped and cost him £25,000. But he was a giant of the Victorian world. A writer in the *Engineer* magazine of the day said that 'he got results, almost, as it were, by instinct… It was quite useless to tell Bessemer that any given device would not answer. He seemed to possess some special power of making things succeed which ought to have failed.' The *Journal of the Iron and Steel Institute*, another contemporary publication, summed his Converter up thus: 'No other invention has had such remarkable results.' He is said to have made £1 million from his patents alone – worth more than £80 million now (RPI) or an astonishing £450 million (average earnings).

The scale of his contribution to the Industrial revolution is clear from the fact that, when he died in 1898, a metal which had cost over £50 a ton in the 1850s could now be bought for *less than £5* a ton. That metal – steel – was absolutely crucial to the development of the railways, shipping and industry, for its ease of manufacture, its quality and its cost. It was men like him – from Savery, Newcomen and Watt onwards – with their incredible drive, creativity and genius, who helped to make Britain, despite its tiny size, for centuries the world's most powerful nation.

THE IPOD

THE insides *and* the outside of **the world's most iconic and revolutionary personal stereo** sprang from British brains.

Thirty years ago, a 23-year-old furniture salesman called **Kane Kramer**, from Hitchin, Hertfordshire, was granted worldwide patents on his 'IXI' music player – a device which stored music on to a computer chip. It held just three and a half minutes of music, but Kramer – who had left school at 15 – rightly believed that technology would one day enable the machine to hold many more tunes. He envisaged a player the size of a credit card, with a rectangular screen and a way of scrolling through tracks.

Sound familiar? Unfortunately, his patents lapsed in 1988 because he couldn't raise the £60,000 needed to renew them. Apple's iPod, unveiled in October 2001, has since sold over 160 million units.

Jonathan Ive, a former Newcastle Polytechnic student, is the man behind the iPod's sleek and elegant look. In contrast to Kramer, who never made a penny from his invention, Ive lives in California, drives an Aston Martin and has received recognition and awards galore for his work. Born in Chingford, Essex, in 1967, he moved to America to work for Apple in 1992. He is now the company's senior vice president of industrial design and the man behind the unique look of the iMac computer, the laptop PowerBook G4, the iPod and the iPhone. In 2008, the *Daily Telegraph* named him **the most influential Briton living in America**, saying that he had 'changed the way Americans – and the world – listen to music and use their mobile phones'.

THE JET ENGINE

THE jet engine is a British invention, and Sir Frank Whittle takes the lion's share of the credit for its development.

But astonishingly **the world's first patent for a gas turbine** was actually granted in *1791* to a Warwickshire pit manager. **John Barber**, from Nuneaton, planned a device to propel horseless carriages, and envisaged blending compressed gas from coal or oil with compressed air and igniting the mixture in an 'exploder', with the escaping gases forced against a paddle wheel. This is the same basic concept employed in today's gas turbines. Sadly, the technology available at the time meant it remained little more than a daydream.

When aircraft came along a century later they relied on propellers for propulsion, until the arrival of **Frank Whittle** (1907-1996). Whittle, from Coventry, knew that if aircraft were to fly at very high speeds and over long ranges, they would have to do so at high altitudes where the thinner air meant less resistance – and where propellers would not work. The answer was the jet engine. His first design was rejected as 'impractical' by the RAF in 1930 but he pressed on, patenting it himself and working on prototypes. He tested **the world's first jet engine** in April 1937 and, having eventually interested the War Ministry, designed the engine for Britain's first jet fighter, the Meteor. This first flew in 1943, but Whittle later said that with Government support he would have provided jet-powered fighters for the Battle of Britain. As it was, his invention made little difference to the war effort. (The German Messerschmitt 262 was actually the world's first operational jet fighter, taking to the air in spring 1944; it, too, had little military impact.)

After the war, on 7 November, 1945 at Herne Bay in Kent, Group Captain H 'Willy' Wilson set **the world's first air speed record by a jet aircraft** when he was timed at 606 mph – faster than any man had previous travelled. Whittle, meanwhile, was promoted to Air Commodore and later knighted. He was awarded £100,000 by the Royal Commission (worth £8 million today) for his invention and, after retiring from the air force, he went to work in the USA, where they eagerly gobbled up his knowledge.

THE JIGSAW

THE world's first jigsaw puzzle was made in London in 1766 to help children learn geography. **John Spilsbury** (1739-1769), a cartographer and engraver from Westminster, stuck a map on to a thin mahogany board and cut it into sections. His plan was to have children reassemble it and learn how countries fitted together. The idea was a success, though his name for it – 'dissected maps' – didn't catch on. The word 'jigsaw' wasn't used until the 1800s and refers to the cutting tool used to make the puzzle. A Spilsbury jigsaw from 1776 can be seen in the British Library.

JURY TRIAL

MODERN jury trials originated in England in 1166 during the reign of Henry II.

A jury comprises 12 members of the public, who hear criminal (and some civil) cases. Many English-speaking countries now have a jury system. It's not perfect – many experts believe that highly complex cases, such as those involving fraud, are beyond juries – but it's better than the systems which it replaced. 'Trial by Battle' involved the accuser 'throwing down the gauntlet' to the accused, and the pair of them duelling to the death. 'Trial by Ordeal' took several forms. In Trial by Water, a suspect was bound hand and foot and thrown into a pond – if he floated he was guilty, if he sank he was innocent. In Trial by Fire, he had to carry a piece of red hot iron for several paces. Three days later the hand was examined; if the burn was infected, he was guilty. 'Evidence' wasn't even an issue – the theory was that God would protect the innocent.

Our Common Law system – where cases are decided by judges using precedents from earlier, similar cases – has also been exported to many countries, including the USA, India and Australia.

THE KALEIDOSCOPE

THE kaleidoscope was invented by a Scottish scientist who was so clever that he went to Edinburgh University aged just 12. **Sir David Brewster** (1781-1868), who was born in Jedburgh, made his first telescope when he was 10. He was interested in optics and the refraction of light, and later went on to create the very large polyzonal lens that was used in lighthouses. But he is best remembered by many for designing his mesmerising children's toy in 1816, using mirrors and loose pieces of coloured glass. Sadly, although he patented the idea, there was a problem with the registration and other entrepreneurs got it to market before him. It was an instant success. Brewster later wrote in a letter to his wife, 'Had I managed my patent rightly, I would have made one hundred thousand pounds by it.'

He named his invention after the Greek words 'kalos' (meaning beautiful), 'eidos' (which means form) and 'scopos' (watcher). So, literally, kaleidoscope means the 'beautiful form watcher'.

KENDAL MINT CAKE

CUMBRIAN confectioner **Joseph Wiper** intended to make clear glacier mints, but he took his eye off the pan and the mixture turned cloudy and grainy. Instead of throwing it away, he tried it and was delighted with the result. It was 1869, and he'd made the first Kendal Mint Cake. His great nephew Robert Wiper later realised its potential as an energy supplier – it's virtually pure sugar. He supplied Sir Ernest Shackleton's 1914-1917 Transarctic Expedition and, in 1953, Sir Edmund Hilary and Sirdar Tensing ate Kendal Mint Cake at more than 29,000ft after becoming the first men to climb Everest.

KEW GARDENS

IT gave the West Indies a staple food, introduced rubber trees to British territories in the east and houses the world's oldest pot plant. What's more – as **the globe's leading botanical research institute** – it may hold the key to saving the planet.

The Royal Botanic Gardens at Kew, which has more than 1.5 million visitors a year, started life as an exotic garden for King George II's wife Queen Caroline in the early 18th century. Had it not been for a brilliant young botanist called **Joseph Banks** it might have stayed that way.

Banks was a gentleman's son born in London in 1743. Passionate about nature, after schooling at Harrow and Eton he accompanied James Cook on his first expedition to Tahiti, the Pacific Islands and New Zealand, bringing back to England hundreds of never-before-seen species, including gardenia, mimosa and eucalyptus. King George III was impressed and asked Banks to become his adviser about matters horticultural, putting him in charge of Kew.

Banks realised that there must be many unknown and exotic plants out in the wider world, just waiting to be discovered, and that many of them might have huge commercial value. This would help drive Britain's Empire-building efforts, so he sent explorers and botanists all over the earth in search of new species – which were brought back in their tens of thousands.

Perhaps his biggest contribution to our modern lives – as well as to the wealth of the nation back then – was to suggest that tea would grow well in India (it did, in abundance – see *Tea*). He also suggested taking breadfruit – a prolific, cauliflower-sized, starchy fruit which can be baked, boiled or fried like potatoes – to the British West Indies from Tahiti. He thought it would make a cheap food for the slaves working in the sugar cane plantations, and it's still a Caribbean staple to this day.

(Incidentally, Banks' efforts to introduce breadfruit to the West Indies led to one of the most infamous maritime events in history, the 'Mutiny on the *Bounty*'. In 1787, Captain William Bligh, a naval officer from Plymouth, set out with a crew of 46 men to Tahiti. After ten tortuous months at sea, they spent a further five rather happier months collecting, potting up and caring for 1,015 breadfruit plants, and getting friendly with the local Polynesian girls. Bligh himself wrote that the women had 'sufficient delicacy to make them admired and beloved', so it's not surprising that many of the sailors didn't want to leave paradise for the cramped conditions on *HMS Bounty*. Once they reached the middle of the Pacific Ocean, acting lieutenant Fletcher Christian led a mutiny and set his captain and 18 crew members adrift in the ship's 23ft long launch. They had very little food and water, but Bligh did have a compass and quadrant and his sailing skills enabled him to make it to Timor in Indonesia – a two-month, 3,500 mile voyage in a dangerously overladen boat, in which only one crewman died. Many of the mutineers were later caught and executed. Banks was furious when he heard about the consequent loss of his cargo, but he arranged for another mission, again led by Bligh, the following year. This time it was successful, and breadfruit were introduced to our Caribbean colonies.)

By the early 1800s, hardly a single ship left any British territory without bringing with it some living or preserved specimen for the Royal Botanic Gardens. One major cash crop was rubber – thousands of seeds were taken from the Brazilian rainforests and replanted in our Malay Peninsula territories in the latter part of that century. These were crucial for the growing rubber industry, as important then as plastics are to us; rubber is still a key part of modern Malaysia's economy today.

In the 21st century, Kew is a major tourist attraction. Among its most famous exhibits is **the world's oldest pot plant**, a 14ft-tall cycad (*E. altensteinii*) which was brought back from South Africa in 1773 by early Kew botanist Francis Masson. It weighed two tons (including soil and pot) when it was last moved in 1980, and was one of more than 1,500 new species, including types of lily and the exotic bird of paradise flower, discovered by Masson.

As to the future, Kew's 'Millennium Seed Bank' is a vast underground vault which will contain seeds from 10% of the world's flowering plants (roughly 30,000 species) by the end of the decade and 25% of them by 2020. The seeds are cleaned, dried and stored in temperatures of -40°C for future use. Every so often the seeds are germinated and distributed to areas of need and eventually it's hoped that every key plant in the world will be represented there, so that if some unknown future event should cause the loss of whole species, replacements can be identified and sown.

THE KNITTING MACHINE

A NOTTINGHAMSHIRE curate invented **the world's first knitting machine** in 1589. Its design was so good it hardly changed for centuries. The **Rev William Lee**, who was born in Calverton around 1550, watched his mother knit, noted how time-consuming a business it was and set his mind to automation. He came up with a hand-powered machine which imitated the movements of the knitter. He believed it would enable paupers to earn more money by speeding up the knitting process, but when he presented his machine to Queen Elizabeth I she was not impressed and refused to issue a patent. According to the *Encyclopaedia Britannica*, she was worried that it would put British knitters out of business and produce more beggars.

Lee's first machine had eight stitches to the inch and produced a coarse woollen fabric. But, with improvements and more needles, it was soon capable of 20 stitches per inch and could knit with silk and produce a finer material perfect for the lucrative silk stocking market. Unable successfully to market his invention in England, he moved with his brother James to Rouen, France, where they produced silk hosiery for French royalty. After William's death, James returned to London, set up at Spitalfields and began producing garments for British gentry. Knitting frames of the early 20th century were almost identical to Lee's first machine.

THE LAWNMOWER

WE invented the ornamental lawn in the 1600s, so it comes as no surprise that we also invented a machine to cut them a couple of hundred years later.

In the early 1800s, lawns were trimmed with a scythe and larger lawns had a line of scythemen with women or children following behind picking up the grass. This was extremely labour intensive (at Blenheim Palace, 50 men worked on the lawns alone) and meant that maintaining a large lawn was a sign of wealth. **Edwin Budding**, born in Stroud, Gloucestershire, in 1795, was an engineer who modified a cloth cutting machine in 1830 to create **the world's first mechanical lawnmower**. Just as with modern machines, the height of the blades could be adjusted and the grass was collected in a box. Being made from wrought iron, it was rather heavy so one man pushed it and a second man pulled it. Budding also invented the adjustable spanner, but unfortunately died of a stroke in 1846.

In the early 1800s, Scottish engineer **Alexander Shanks** developed a larger version that was pulled by a pony, and steered by an operator walking behind it. Incidentally, the world's premier (admittedly the only) lawnmower museum can be visited in the north west seaside resort of Southport. Best to go on a rainy day.

LIFEBOATS

THE battle over who invented the lifeboat is being fought from beyond the grave. The only certainty is that the designer was British: three Britons make the claim, and two of them have it engraved on their tombstones.

The gravestone of **Lionel Lukin**, an Essex coach maker who was born in 1742 and died in 1834, is engraved with the following inscription: 'This Lionel Lukin was the first who built a life-boat, and was the original inventor of that principal of safety by which many lives and property have been preserved from shipwreck.'

The gravestone of his rival, South Shields parish clerk **William Wouldhave** (1751-1821), reads: 'Inventor of that invaluable blessing to mankind the Lifeboat.'

The third man to make the claim is **Henry Greathead**, a contemporary of both men who was born in Richmond, Yorkshire, in 1757 and died in 1816.

Lukin obtained a patent for his 'unsinkable' boat in 1785. It was for a vessel that would 'neither overset in violent gales or sudden bursts of wind, nor sink if by any accident filled with water'. He tested it on the River Thames and the first, called *Experiment,* was taken to Ramsgate where it may have been used for smuggling rather than life-saving. In 1790, Lukin – who also invented a rain gauge and an adjustable reclining bed for patients in hospitals – published scale-drawings and a description of his lifeboat. In total, only four were ever ordered.

In the same year as Lukin received his lifeboat patent, Henry Greathead moved to South Shields and started work as a boat builder. In 1789 there was a tragic shipwreck close to his home when a vessel called *Adventure* sank with all hands in a storm. A competition for the best lifeboat design was launched and Greathead won. Lukin complained that it infringed his patent but the *Original,* a 30ft long, cork-lined, 12-oared boat sailed for 40 years before finally being wrecked against some rocks, killing two men.

Wouldhave, who was described in the *Journal of the National Shipwreck Institution* in September 1852 as 'a clever but wayward man', also entered the competition and later claimed that *he* designed the boat that Greathead built. He said he came up with the design after watching a broken wooden dish floating on a bowl of water. No matter how many times he tried to capsize it, it always righted itself.

LINOLEUM

WHEN **Frederick Walton** left the lid off a tin of paint he found that a rubbery skin had formed on the top. It led to the invention of 'linoleum', which has since protected millions of square feet of floors all over the world.

Walton, who was born in Halifax, Yorkshire, in 1834, determined that the skin was formed by oxidised linseed oil. He made the first lino from linseed oil mixed with cork dust and ground limestone. It was painted onto a backing material, allowed to dry and built up in layers. He patented his hard-wearing product in 1860, and four years later he set up the Linoleum Manufacturing Company in Staines, Middlesex. He was soon exporting to Europe and the United States, and set up the American Linoleum Company on Staten Island, New York, in 1872. Vinyl flooring – often erroneously called linoleum – became more popular in the 1970s, but Walton's invention is gaining popularity again as it is organic, biodegradable and recommended for people with allergies.

LIQUORICE ALLSORTS

GEORGE **Bassett** set up a sweet-making factory in Sheffield in 1842, but it took a happy accident and more than 50 years for the arrival of his most famous confection, Liquorice Allsorts. A Bassett's salesman called Charlie Thompson visited a wholesaler with a selection of sweets in 1899. The customer turned his nose up at each item and, as Thompson gathered his samples boxes together to leave, he knocked them over, spilling the sweets on the counter. The wholesaler's eyes lit up, and he immediately placed an order for the *mixed* sweets. Thompson named them Liquorice Allsorts. Bertie Bassett made his first appearance in 1929, and 14 million individual sweets are now produced daily. Jelly Babies are another Bassett invention. Originally known as 'Peace Babies', they were launched to celebrate the end of World War I.

THE LONGITUDE PROBLEM

FEW people have heard of Sir Cloudesley Shovell despite his unforgettable name and his undoubted status as a true British war hero.

Born in 1650, Shovell was a great sailor and warrior who led his men to victory in many battles against the French and Spanish, and had the run of the Atlantic Ocean and the Mediterranean during the late 17th and early 18th centuries. He had joined the Navy as a cabin boy. At the age of 13 he was serving in the West Indies, and 40 years later he was still at sea, by then an Admiral commanding a fleet of warships. He was known for his seamanship and his 'unparalleled courage, capacity, and honesty'; it's fair to say that if *anyone* knew sailing, Cloudesley Shovell did.

With all his great experience and ability, then, it comes as something of a surprise to learn that he drowned in 1707, along with 1,314 of his men, in one of the worst disasters in our maritime history. Even more surprisingly, it happened not in some far-off sea but on one foggy evening in our own well-charted waters.

How could one of our nation's most senior and skilled seamen allow three of his ships to run aground and sink with such catastrophic results?

The answer is simple – he thought he was entering the mouth of the English Channel, when in fact he had missed it by miles and was actually heading straight for the rocks around the Scilly Isles, some distance off the tip of Cornwall.

Why? Because, in common with every other sailor in the world in his day, once he had lost sight of land he could not say with certainty where he was.

This was because there was no way of establishing one's longitudinal position. A Briton found a way, and **revolutionised global travel** in the process.

For navigational purposes, the globe is divided up into sections delineated by two sets of invisible lines, each running at right angles to the other. One set – lines of latitude – travel from east to west. The other set – lines of longitude – travel from north to south. If you know which line of latitude and which of longitude you are on, and you have adequate maps and charts, you can pinpoint your exact position anywhere on the surface of the globe.

Latitude was relatively easy. The equator provides a natural centre line, and all the other lines of latitude run parallel to it. By measuring the angle of the Sun to the horizon at midday, when it is at its highest point, and comparing the result with a book of daily tables giving the Sun's position vis-à-vis the equator, it was possible to establish one's latitudinal position. (At night, the angle of the Pole Star to the horizon was used instead. A quadrant, sextant or octant was used to make the measurements – the first two of these devices having been invented by Britons.)

Longitude was a different matter, because in the north-south plane there was no fixed starting point like the equator against which to measure the position of the Sun.

To get around this problem, early navigators used 'dead reckoning', which meant working out where you were by calculating your speed and course since leaving your last *known* position. Speed was estimated by using a 'chip log', a knotted rope tied to a wooden log which was thrown from the back of the ship, the knots being counted and timed as they passed through a sailor's hands. Done continuously, it gave some clue as to the distance travelled. Meanwhile, the course was worked out with a compass. But both of these methods were fraught with error: sailors miscounted the chip log, and compasses were affected by magnetic variations. The result was that longitudinal position was often little more than an educated guess.

The only way to know for sure where you were was to sail down (or up) to the latitude of your destination and then keep to that latitude until you arrived. But this made it easier for pirates to find their prey, and also dramatically lengthened the distance to be sailed and the time taken, increasing the cost of voyages and the risks of death from scurvy or even starvation.

After the Cloudesley Shovell tragedy, and with oceanic trade becoming ever more important, the government announced a prize of £20,000 – £30 million in today's terms – for anyone who could crack the longitude problem. Predictably, the huge sum of money attracted scores of unfeasible solutions, including anchoring a chain of ships every seven miles across the sea to act as markers.

There was a more practical and scientific method available, however, and it related to Time.

The Earth rotates on its north-south axis through 360 degrees in one day, or 15 degrees per hour (without getting too complex, each degree is further sub-divided into 'minutes' and 'seconds').

As every modern-day international air traveller knows, this means that when it is noon in London it is midnight on the opposite side of the world – which is 180 degrees away. Equally, it is 6am 90 degrees (or a quarter of the world) away to the west, 6pm 90 degrees away to the east, and every other hour of the day (and division thereof) at all the different longitudinal positions on the earth's surface in between.

This means that if you fix the longitudinal line which passes down through London as 'zero' then you can work out your longitudinal position by comparing your local time with the time in London.

If you add your latitudinal position, you now know *exactly* where you are.

For example, imagine you have sailed due west into the Atlantic Ocean on exactly the same latitude as London – 51.40 degrees. If it is 9am local time, and noon in London, you must be at a longitude of exactly minus 45 degrees (15 degrees further west for every hour's difference). Combining the two puts you at 51.40.00/-45.00.00, some 300 miles off the coast of Newfoundland.

So far, so simple (relatively speaking).

However, while it was easy to work out local time from the angle of the Sun relative to the horizon, the difficulty lay in knowing what time it was in London, thousands of miles away over the distant horizon.

Again, some ludicrous suggestions were made. Perhaps the most barking, if you'll pardon the pun, involved the 'wounded dog theory'. 'Digby's Powder' was a supposedly 'magickal' substance which cured wounds remotely if applied to something belonging to an injured person. The theory called for an injured dog to be taken aboard a ship. After it sailed, back in England someone would dip a bandage from the dog's wound in Digby's Powder every day at noon. Many miles away, aboard the ship, the dog would supposedly yelp – the powder was painful when applied – and the captain would know that it was midday in London. It was quite bonkers, of course, and utterly useless as an aid to navigation. Nowadays, it would also be banned by the RSPCA.

The obvious solution was rather more prosaic: take a clock with you, set to London time.

Again, this was easier said than done. It needed to be highly accurate, because an inaccurate clock was no different to a chip log sailor who couldn't count. Clocks *did* exist which kept good time, but that was on land. On a ship, rolling in every direction, pounded by waves, wind and storms, constantly damp, corroded by seawater and salt air, facing the huge changes in temperature and humidity involved in sailing from Plymouth to Jamaica – well, it's easy to see how even a genius like Sir Isaac Newton doubted it was possible to manufacture a timepiece able to cope with all of that.

That would probably put most people off, but fortunately **John Harrison** was not like most people. He ended up devoting 40 years of his life to the job (the full story of which is beautifully told in the bestselling *Longitude* by Dava Sobel).

Harrison was born near Wakefield in Yorkshire in 1693. Like his father, he became a carpenter and at the age of 20 started making wooden clocks (one of which is now to be seen at the Science Museum). He steadily worked on improving his designs and by the mid-1720s he had built a clock which lost no more than a second per month.

By 1730, he felt confident enough to go to London and present his designs for a sea-going version to the astronomer royal, Edmond

Halley, hoping that he would help fund the construction. Halley sent Harrison to visit the eminent clockmaker George Graham, and Graham, intrigued by the blueprints, paid for the first ship's chronometer to be built.

Called 'H1' (for 'Harrison 1'), it weighed 72lb, was two feet tall and took five years to put together. Its workings compensated for the movements of a ship, moisture and temperature changes, and even ran without having to be oiled. When Harrison took it to London, its complexity and obvious quality made it the talk of the scientific community. H1 was given a sea trial in 1736 – a voyage to Lisbon on a Royal Navy warship. On his return, the Captain of *HMS Orford* praised the clock and admitted that it had allowed for longitudinal calculations far more accurate than those he was himself able to produce by other methods.

That was merely the beginning. Over the next 24 years, Harrison refined his designs, producing H2 and H3, before arriving at his masterpiece, H4. This was a magnificent piece of work, with a diameter of just over five inches and a weight of only 3lb. It was sent on two sea trials. The second, a voyage to Barbados in 1764, took 47 days, during which the chronometer lost just 39.2 seconds, allowing for astonishingly accurate navigational calculations – these were only 9.8 miles out after a journey of some 5,000 miles. This should have won him the £20,000 prize on the spot, but the Board of Longitude, which had been set up in 1714 to oversee the search for a marine clock, refused to hand over all of the money because they could not believe its accuracy. Grudgingly, in 1765, they paid Harrison half of the money and had another watchmaker, Larcum Kendall take away H4, dismantle it and copy it. (His 'K1', finished in 1769, was itself a fabulous piece of engineering which went with Captain James Cook on his second and third voyages.)

Meanwhile, Harrison, furious that the Board of Longitude had refused to pay him all of the £20,000 they surely owed, retired to his workshop and built yet *another* clock, H5. He unveiled it in 1770 and a 10-week trial at the Royal Observatory in Kew saw it lose less than a third of a second per day. Still the Board of Longitude would not hand over the remaining £10,000. In fact, it was only in 1773, after Harrison appealed to King George III and the Prime Minister, Lord North, that Parliament agreed to give him a further £8,750 in recognition of his genius.

On the upside, he was now the equivalent of a modern millionaire. On the downside, he was 80 years old, so his enjoyment of his richly-deserved wealth was limited, both in scope and in time. In fact, he died three years later.

'H1' to 'H4' are now on display at the National Maritime Museum in Greenwich, and 'H5' can be seen at the Clockmakers Museum in London.

A footnote: astronomers believed that 'The Method of Lunar Distances' offered a better way of calculating longitude. This involved observing the angles and distances between the Moon and stars as the former moved across the face of the night sky, and cross-referring with a complicated almanac of tables to pinpoint time and position. Although it was far better than dead reckoning, it was harder to do and less accurate than using a chronometer, and while it had its supporters it died out as decent clocks became cheaper and more widely available. Incidentally, thanks to the work of Harrison and his lunar method rivals, the Greenwich meridian – which gives a longitude of zero degrees – was internationally adopted in the International Meridian Conference of 1884. Greenwich Mean Time is now used worldwide as a standard time irrespective of location; everywhere is X hours 'ahead of' or 'behind' GMT.

THE MACHINE GUN

LONDON lawyer **James Puckle** invented **the world's first machine gun** at a time when the average soldier was still firing muskets.

The Puckle Gun was granted a patent in May, 1718. It was tripod-mounted, had a 3ft barrel and 1.25in bore and a revolving cylinder which fed bullets into the gun's chamber. It was able to fire nine shots per minute – a musket could be fired and reloaded only three times in the same period – and worked in the rain, unlike other contemporary weapons. Puckle designed it so that it could fire conventional spherical bullets of the time and special cuboid ammunition for use on Turks. He believed, correctly, that cuboid bullets would inflict more harm. Perhaps more optimistically, he also believed they would convince the Muslim Turks of the 'benefits of Christian civilisation'.

The *London Journal* reported that at a demonstration of the weapon 'one man discharged it 63 times in seven minutes, though all the while raining'. Unfortunately, the British Army was not interested. A newspaper report from the time said that the only people wounded by the gun were 'those… who hold shares therein'.

The world's first modern machine gun was also a British invention (though its originator **Sir Hiram Maxim** was a naturalised Briton, having been born in America). The Maxim Gun, launched in 1885, was a water-cooled, belt-fed weapon which could fire 500 rounds per minute (when a trained rifleman could fire 20). Maxim tested it in his London garden – nowadays, that probably wouldn't be allowed – and eventually went deaf from the noise. He had been moved to design the weapon after meeting a man who told him, 'If you want to make a pile of money, invent something that will enable these Europeans to cut each others' throats with greater facility.'

Maxim's firm later merged with Vickers and his machine gun metamorphosised into the Vickers gun, the standard British Army weapon for many years.

THE MACKINTOSH

A SCOTTISH chemist's name was immortalised (though slightly misspelled) when he invented **the world's first completely waterproof fabric**.

Charles Macintosh, a dye factory owner's son born in Glasgow in 1766, had a passion for science. While experimenting, he found that if he dissolved India rubber in coal tar naphtha and painted the solution between two pieces of cloth, the material was impervious to water once dry. He patented his fabric in 1823. The Mackintosh (with the addition of the letter 'k') sold well, but was not without teething troubles. The early coats were stiff, smelly and not very comfortable, and the rubber bonding could disintegrate in cold weather or melt in heat. 'Macs' became less fashionable with the introduction of the railways, as people were no longer as exposed to the elements as they had been in stagecoaches or on horses, though the material was still in demand for use in tents and nautical equipment made for the intrepid explorers of the day. The coats improved when Macintosh went into business with Mr Elastic, Thomas Hancock (see *Elastic*). The sheets of rubber produced by Hancock's 'Pickling machine' (a rubber masticator) worked better than early rubbers.

MAIL ORDER

A WELSHMAN set up **the world's first international mail order business**.

Pryce Pryce-Jones, of Newtown, Montgomeryshire, was born in 1834, and started work as a shop assistant, aged 12. A solicitor's son, he founded his own business, Royal Warehouse (later named Royal Welsh Warehouse), in 1859. Based in his hometown, it was a small shop selling Welsh flannel and woollen goods. His moment of inspiration came when he realised he could sell more if he could reach people who couldn't visit his shop. He sent samples of flannel and price lists to local gentry, and had their orders delivered via stage coach. The expansion of the railways and a more efficient postal service soon meant he could reach a wider audience still.

Leaflets were printed and distributed, people selected items they wanted and he posted them out. He used the names of wealthy or distinguished customers for endorsements – among them were

Florence Nightingale and various European royals, including Queen Victoria to whom he supplied woollen undergarments. According to the *Oxford Dictionary of National Biography*, by 1890 he had 200,000 customers around the world. By 1892, there was a post office inside his warehouse to speed up dispatch.

Americans claim that New Jersey travelling salesman Aaron Montgomery Ward invented mail order. But he did not produce his first catalogue – a one page list of 162 items – until 1872. Northumberland businessman David Landreth sent out a seed catalogue in 1784, and Benjamin Franklin offered scientific books for sale in Philadelphia as early as 1744, but these were very small in scale: Pryce Pryce-Jones went global.

MALARIA

THERE are up to 500 million new malaria cases a year and an estimated million people now die of the disease annually. Some 200 million are thought to have been killed through human history, Alexander the Great and Genghis Khan being two of many famous figures among its victims. But it was not until 1898, following the painstaking work of British Army captain's son **Ronald Ross**, that the cause of malaria – parasite-infected mosquitoes – was established.

Victims suffer chills and fever, headaches, vomiting and diarrhoea. In severe cases, it can attack the brain and central nervous system and cause coma, paralysis or death. For thousands of years, it remained a mystery, called 'malaria' – literally, 'bad air' – to reflect the belief that the infective agent was somehow carried on the winds. But in 1880, the French doctor Charles Laveran found that infected blood cells carried a parasite. This discovery, for which he was awarded a Nobel Prize, explained what caused malaria. However, it did not explain how the parasite – a protozoan called Plasmodium – entered the bloodstream in the first place. Ross (1857–1932), who worked as a doctor in India, was determined to understand this process. Eventually, after years of meticulous research, during which he caught the disease himself, he found Plasmodium in the saliva glands of a mosquito that he had artificially fed on a patient who had the disease. Then he demonstrated that malaria could be transmitted from infected birds to healthy ones by a mosquito bite. It was a short hop to prove that malaria could be passed on to humans in the same way.

These were huge breakthroughs because they offered the first glimpse of ways to defeat malaria. Although effective medicines were few and far between, Ross's work – he, too, won a Nobel Prize – meant that a programme of eradication aimed at the insect hosts could begin. In Britain and elsewhere, 'mosquito brigades' of volunteers began eliminating mosquito larvae from the stagnant pools and marshes where they breed. Thanks to this and drainage of wetlands, malaria is now not found in the UK.

Scientists are always working on cures, of course. Excitement currently centres on extracts from a fern-like plant with yellow flowers called wormwood (*Artemisia annua*); the Bill and Melinda Gates Foundation recently gave $13.6m to York University, where cutting edge research is underway into the development of high-yield strains of artemisia.

THE MARATHON

WE didn't invent the marathon, obviously. The original was supposedly run by a soldier, Pheidippides, carrying a message from the battle of Marathon back to Athens in 490 BC, and the first modern marathon was conceived by the French and run at the Athens Olympics in 1896. However, early marathons were not run over fixed distances, and the length of the race – 26 miles and 385 yards (42.195km) – was set by the British at the London Olympics of 1908.

MARMALADE

LEGEND has it that marmalade was first made in the late 1700s by a Scotswoman confronted with a ship-load of inedibly bitter oranges. James Keiller, a grocer born in Seagate in 1775, bought Seville oranges from a Spanish ship sheltering in Dundee harbour from a storm. But no-one wanted the sour, pip-ridden fruit, so – not wishing to waste them, sweets and jam-maker **Janet Keiller** (some say she was James' wife, others that she was his mother) set to work with them and came up with marmalade.

The word marmalade comes from the Portuguese word 'marmelada' – a solid, grainy jam made from cooked quinces and sugar – and it may be that Mrs Keiller simply adapted another's recipe. But the Keillers were certainly the first to produce it and sell it on a mass scale. They opened the first marmalade factory in Seagate in 1797. By the early 1800s, James Keiller and Son Dundee Orange Marmalade was being sold as far afield as Australia, New Zealand, South Africa and China.

MATCHES

AROUND 500 *billion* matches are produced each year. They're a British invention, the brainchild of a dapper little man who was known in his home town as the 'Stockton Encyclopaedia'.

John Walker (1781–1859), a grocer's son, trained as a surgeon but found it too gruesome. In 1819, he set up a chemist's shop in Stockton High Street, where he was well known for being a font of knowledge and for wearing a tall beaverskin hat and a white cravat. While experimenting in his laboratory, he dipped the end of a stick into a batch of chemicals – mainly potassium chlorate and antimony sulphide. He found that once the chemicals had dried he could strike the stick anywhere and make a flame (safety matches were developed in Sweden later so that you couldn't make a flame 'anywhere').

The first recorded sale of Walker's 'Friction Lights' was to a Mr Hixon, a Stockton solicitor, on April, 7, 1827. They were sold in batches of 100 for a shilling, and a piece of sandpaper (for striking – and also a British invention) was supplied with each sale. Walker, a bachelor, never patented his product, and it was copied by Samuel Jones who ran a shop called The Lighthouse in The Strand in London. He named his matches 'Lucifers' and started selling them in 1829.

MECCANO

IN 1901, a Liverpool shipping clerk was granted a patent for a construction toy called Mechanics Made Easy. Six years later, its name was changed to Meccano and sets containing reusable metal strips, screws, bolts, nuts, wheels, axles and rods, were being sold to small boys worldwide. By 1928 its inventor, **Frank Hornby**, employed more than 1,200 people at his Liverpool factory. He was also the brains behind Dinky cars – toy vehicles which are now highly prized by collectors who pay hundreds of pounds for rare examples – and Hornby train sets. When he died in 1936 he left £230,000 – worth more than £42 million today.

MICROWAVE OVENS

MICROWAVES – electromagnetic waves which lie between infra red and radio waves in the electromagnetic spectrum – were predicted by the Scottish mathematician and theoretical physicist **James Clerk Maxwell** (see *Electromagnetism* and *Photography)* in 1864. Today, they have many applications. Mostly, these are in ultra high-tech fields like communications and satellite navigation – but they're also brilliant for heating up a left-over curry.

The microwave oven would not be possible without the work of two brilliant Britons, **Dr John Randall** (1905-1984), from Newton-le-Willows, Lancashire and **Dr Harry Boot** (1917- 1983), a physicist who was born in Birmingham. They worked on a key component of the microwave oven, the cavity magnetron, at Birmingham University during WWII. This was nothing to do with curry – the cavity magnetron, a vacuum tube which produces the microwaves, is also a key component of radar. (We won't go into exactly what one *is*, though. It would take several pages, and no-one without a PhD in physics would understand it, including me.) An American colleague noticed that his chocolate bar melted when left near the magnetron and the idea for a new oven was born.

THE MOON

THE illustrious Italian astronomer Galileo Galilei generally takes the credit for being the first to view the Moon through a telescope – but a Briton, Oxfordshire cartographer and mathematician **Thomas Harriot** (1560-1621), beat him to it. Aged 49, and then living in Isleworth in London, Harriot used a refracting telescope (see *Telescopes*) to view the satellite on July 26, 1609 – at least four months before Galileo. Until then, there was debate as to whether it was even spherical. Harriot saw that it was, and also the craters, created by falling meteorites, which pitted its surface and offered compelling evidence of the universe's violent and ever-changing nature. He produced **the world's first detailed drawings of the Moon**, and observed comets and the moons of Jupiter through his various 'spyglasses'. But he never published his work, which goes some way towards explaining why Galileo was able to claim his glory.

Harriot's first telescope had a magnification of 6x, but by April 1611 he was working with a 32x magnification. The lenses were made by his servant, Christopher Tooke. Sir Walter Raleigh employed Harriot as a scientist – due to his expertise in map-drawing and navigation – and he sailed with him to America in 1585. (Some credit Harriot with introducing potatoes to the British diet.) Later he worked for William Percy, ninth earl of Northumberland, and was briefly imprisoned along with Percy on suspicion of links to the Gunpowder Plot. He also made a huge contribution to the understanding of algebra and refraction.

A pipe-smoker, probably due to his early friendship with Raleigh, Harriot died aged 61 from a cancerous tumour in his nose, which also gives him the unfortunate distinction of being one of the first people to die from a smoking-related disease. A crater on the moon was named after him in 1970.

THE MOUSETRAP

LEEDS ironmonger **James Henry Atkinson** invented the 'Little Nipper' – **the world's first successful mousetrap** – in 1899. It's a wooden block with a metal bar which is closed by a powerful spring activated by the mouse stepping on a pressure plate to grab whatever bait is placed in the trap. The bar closes with tremendous speed, killing the unfortunate rodent instantly. Atkinson sold the rights to a company called Procter Brothers for £2,000 in 1913, and they continue to make the traps to this day at their factory in Bedwas, Caerphilly. Despite many high-tech rivals, the Little Nipper has never been bettered and still holds 60% of the market.

MRI SCANNERS

MORE than 60 million MRI scans are performed in hospitals each year, giving doctors a vital picture of the internal workings of the human body. But without the vision of a British physicist, Magnetic Resonance Imaging might not have become a routine medical procedure.

Like CT scans, they eliminate the need for investigative surgery. But unlike CT scans, which use x-rays, MRI scanners work by flooding the body with strong magnetic and radio waves – believed to be harmless.

. Computer programmes then analyse the waves and convert this information into detailed images of internal organs, joints and other tissues. **Sir Peter Mansfield**, a scientist working at Nottingham University, came up with a mathematical method of deciphering these signals and quickly turning them into 3D computer images. This made MRI a practical diagnostic tool. Mansfield, a keen helicopter pilot who was born in London in 1933, was the first human guinea pig for the initial full-body machine in the 1970s – with his wife on hand in case anything went wrong. It didn't, and in 2003 his breakthrough led to him sharing a Nobel Prize with an American MRI researcher, Professor Paul Lauterbur – not bad for a man who had left school, aged 15, without a single O level, to become a printer's assistant.

SIR ISAAC NEWTON

MOST of us, if asked to name the greatest genius the world has yet produced, would probably suggest Albert Einstein. The difficulty would come if we were then asked *why* we had suggested him. After all, you could probably fit everyone who truly understands his Theory of Relativity onto a double decker bus. So who do those people who *are* conversant with Einstein's work say possessed the most brilliant mind humanity has seen? Interestingly, most actually name a Briton – **Sir Isaac Newton**.

The great Russian Lev Landau, for instance, was a Nobel Prize winner who discovered and proved things in scientific disciplines many of us can scarcely pronounce – such as quantum electrodynamics, quantum mechanics and plasma physics. He famously drew up a list of the greats of physics which ranked Newton top, with Einstein second. In 2005, a survey of 345 scientists by the Royal Society emphatically agreed with Landau: asked who had made the bigger overall contribution to science, 86.2% said Newton and 13.8% Einstein.

Isaac Newton was born on Christmas Day 1642*, in the Lincolnshire hamlet of Woolsthorpe-by-Colsterworth, just off the modern A1 a few miles south of Grantham. His father, a yeoman farmer who was unable to sign his own name, had died three months earlier, and Isaac was born prematurely. He survived this, obviously, and within a few years his mother Hannah had remarried, a Reverend

---

* Or 4 January 1643, as we now know it. Pub quiz question: What date followed Wednesday September 2, 1752 in Great Britain? Answer: Thursday September 14. We had used the Julian Calendar, but it had gradually drifted ten days out of sync with the seasons, so we changed to the Gregorian Calendar in September 1752. Many people were unhappy about this, believing that the number of days that their lives had been allocated was fixed, and that they had been robbed of ten days. Other countries gradually followed suit, though Greece didn't change until 1923.

Barnabus Smith. (The young Newton did not much like the Rev Smith; Natalie Rosinsky's biography talks of him having threatened to burn his mother and stepfather 'and the house over them'.)

He was educated from the age of about twelve at The King's School in Grantham where, as bored youths often do, he carved his name into a windowsill where it can still be seen, and then Cambridge. His early years there were interrupted by the Great Plague of 1665-6, which killed around 100,000 people; the university was closed and he returned to Lincolnshire. After the plague subsided, he went back to Cambridge and in 1669 was appointed Lucasian Professor of Mathematics.

His interests and work were wide-ranging (and we can do no more than touch on them here). He lectured in Optics for three years, constructing **the first practical reflecting telescope** (see *Telescopes*), and made huge advances in the **understanding of light** – showing that white light could be split or refracted into a spectrum of colours using a prism, and then reformed using another. He suggested that light appeared to behave sometimes as though it was a wave and at other times a particle (he was right, but this idea was not taken seriously for another 250 years).

In Mathematics, his most important contribution was the **invention of 'calculus'**; this allowed accurate calculations of physical processes, such as the motion of a projectile, for the first time. It is used today in every field of science and engineering, from working out the flow of water around a ship to the stresses on the bridge it sails under.

In his greatest work, The *Philosophiae Naturalis Principia Mathematica* published on July 5, 1687, he produced three **Laws of Motion** and his **Law of Gravity**. The former (the most famous of which states that, effectively, 'Every action has an equal and opposite reaction') explain how forces act on given bodies. The latter describes how everything – even a grain of sand – has gravity, but that this force varies between objects depending on their mass. The greater the mass of an object, the greater its gravity. Taken together, these four laws explained how the solar system worked – that is, with the Sun at the centre and the planets orbiting around it – and finally put paid to any idea that the Earth was at the centre of the Universe.

Of course, Newton is best known for his explanation of gravity. His moment of clarity came as he walked through a garden in Lincolnshire in 1666. Many years later, his assistant John Conduitt wrote that 'it came into his thought that the power of gravity (which brought an apple from a tree to the ground) was not limited to a certain distance from earth, but that this power must extend much further than was usually thought. Why not as high as the Moon said he to himself & if so, that must influence her motion & perhaps retain her in her orbit, whereupon he fell a calculating what would be the effect of that supposition.'

Newton's calculations of the force of gravity were later improved by Einstein, but his laws still work well enough to be used by NASA and other space agencies today when calculating the trajectories of satellites and probes.

This would be an impressive output for half a dozen people, but it represents only a fraction of Newton's achievements. Among many other things, he also constructed a static **electricity generator**, developed equations to calculate the **speed of sound** and others to work out the rate at which hot objects cool. It's fair to say that he did more than any other scientist in history: despite this, much of his work was not really appreciated until long after his death, perhaps because he was so far ahead of most other people that they could not understand its full significance. His knighthood was awarded not for being the cleverest man the world had ever seen but probably for his enthusiastic pursuit of counterfeiters after being made Master of the Royal Mint in 1699. (Despite being Master of the Mint and an all-round genius, Newton did lose a fortune in the South Sea Bubble, a speculative boom-turned-to-bust of his day.) He was also briefly an MP from 1689, but his only recorded words in Parliament seem to have been a request to close the window because of a cold draught that was bothering him.

As for the eternal Newton vs Einstein debate, it seems likely Newton himself would have dismissed it as nonsense. Like most (if not all) of the other inventors, artists and scientists in this book, he had built on brilliant work by other people, and he knew it. In a letter to his contemporary **Robert Hooke** in 1676, he acknowledged the achievements of previous greats, writing: **'If I have seen further it is because I have stood on the shoulders of giants.'**

(Incidentally, Hooke himself possessed one of Britain's finest and most wide-ranging minds. A clergyman's son born on the Isle of Wight in 1635, he is best known for his **Law of Elasticity**, published in 1660 as 'ceiiinosssttuv', an anagram for the Latin phrase 'ut tensio sic vis', or 'as the extension increases, so does the force'. This might seem a rather bizarre way of announcing your findings to the world, but it was used by many scientists at the time so that they could prove they'd thought of an idea first, without giving it away immediately to rivals if it needed further testing or refinement. Hooke's law didn't need any refinement, and led on to the development of the balance spring, which allowed the manufacture of the wristwatch. He also invented or improved the barometer, anemometer [which measures wind speed] and hygrometer [an instrument used for measuring humidity], made major contributions to astronomy [he was the first person to note the rotation of another planet – Jupiter, by watching its red spot move over several nights] and helped Christopher Wren on his redesign of London after the Great Fire of 1666. Working with Robert Boyle, a famous Irish scientist, he created an air pump which could create a partial vacuum, and together they showed how air could be compressed. They also discovered that fire was extinguished if air was pumped out of the chamber, but that gunpowder would still explode. In 1671, Hooke built a sealed chamber and sat in it as an assistant furiously worked the air pump. Thus, he became the first man to experience altitude sickness at sea level: leaks in the chamber caused by the deficiencies of 17th century engineering probably saved his life. He produced the first useful '**universal joint**', an important part of every car [it allows an otherwise rigid rod to move at any angle, and is important in drive shafts], and made major breakthroughs in microscopy. The microscope had only been invented 30 years previously, and Hooke made many early observations, examining the structure of wood and ice crystals and analysing the beating action of insect wings, hoping to gain some clue as to how to copy them. He coined the word '**cell**' – he thought the building blocks of the organic particles he examined resembled the chambers, or cells, in which monks worked – and his detailed drawings were published in his book *Micrographia* in 1665, which sold for a whopping £1.50 a copy, or about £350 in today's money. Samuel Pepys described it as the most fascinating book he had ever read. Hooke never married, and amassed a large sum of money – I cannot say whether the two facts are related. Unlike every other famous person of the era, no portrait exists of him. He is buried somewhere in North London; if his remains are ever found, there are plans to use laser facial reconstruction techniques to produce an image of him. This would be a fitting tribute to someone described as 'The Man Who Knew Everything'.)

Back to Sir Isaac Newton: the great man died in 1727 and was buried in Westminster Abbey. The poet Alexander Pope summed up his achievements thus:

'Nature and Nature's Law lay hid in night
God said "Let Newton be" and all was light.'

Einstein said of him, 'In one person he combined the experimenter, the theorist, the mechanic and the artist in exposition.'

The great French mathematician Pierre Laplace said, 'There is only one universe, and it has only one set of laws, and there will only ever be one man who discovered those laws.'

There is a bust of Newton in Leicester Square: it is falling apart.

NUCLEAR POWER

IT'S a bit of a bogeyman for many people, but once the world's coal and oil runs out we may be very grateful for the development of nuclear power. No one nation can lay claim to having 'invented' it, but we did build **the world's first commercial nuclear power station**. Calder Hall was erected near the village of Seascale on the Cumbrian coast (Sellafield is next door), and connected to the National Grid in August 1956. It had four reactors, each of which could generate fifty megawatts of power (a megawatt is one million watts, so 50mw is enough to light half a million 100-watt bulbs), and operated until March 2003.

The man who oversaw the development of Calder Hall was **Sir John Cockcroft**. Born in 1897 to a family of weavers and mill owners in Todmorden, Lancashire, Cockcroft won the Nobel Prize (shared with the Irish scientist Ernest Walton) for '**splitting the atom**' – demonstrating how atoms could be broken up by bombarding them with high energy particles. This was the first step towards the process which provides power in a reactor.

Cockcroft and other British Nobel Prize-winners, like the Mancunian physicist **Sir Joseph Thompson** (who in 1897 discovered the existence of **electrons**, the particles which orbit the nucleus, or centre, of an atom) and Cheshire scientist **Sir James Chadwick** (who in 1932 discovered the **neutron**, which along with the proton, makes up the nucleus of an atom) were at the very forefront of nuclear physics. The basis of the science had been laid a century earlier by the British scientist **John Dalton**. Dalton, a Quaker from Cockermouth, first theorised in the early 1800s that each element was made up of atoms, every one of which was identical and unique to that element. (He also identified the existence of colour blindness for the first time.)

Our current nuclear reactors rely on the *fission* (breaking up) of uranium which unfortunately creates radioactive waste. The 'holy grail' of energy is nuclear *fusion* – a much cleaner process which involves *combining* lighter elements such as hydrogen. This powers the Sun: if we could replicate it on Earth, we'd have a permanent source of clean power. The Joint European Torus (JET) facility in Oxfordshire is **the world's leading fusion research centre**, and has been working on building a fusion reactor since 1983; so far it remains a far-off dream.

NURSING

FLORENCE Nightingale's best known achievements are in improving nursing – but she was also a gifted mathematician who invented the **pie chart**.

Born into a wealthy family during their 'tour' of Europe in 1820, she was named after the city of her birth. Educated by her father William at their homes in Derbyshire and Hampshire, she developed a love of mathematics, particularly statistics, and shunned the traditional feminine pursuits of the day. Despite opposition from her parents, she decided to devote her life to nursing. Her first job was as unpaid superintendent – her father had given her an income of £500 a year (£35,000 a year today) – at The Establishment for Gentlewomen During Illness in Harley Street, London.

In 1854, the Crimean War – between Russia and Britain, France and the Ottomans – began, and articles soon appeared in *The Times* describing the appalling conditions in military hospitals. Nightingale travelled to the Crimea with a party of 38 female volunteers and was horrified by what she found. There was a shortage of medicines and food, and the injured men were lying in squalor, surrounded by vermin. Typhus, cholera and other infections were rife. She immediately set to work improving life for the wounded men.

She is said to have treated 2,000 patients herself – she became known as 'The Lady With The Lamp' by the troops – but she also collected data on the mortality rate, and calculated that soldiers were seven times more likely to die from disease in hospital than on the battlefield. She improved sanitary conditions, moved beds further apart, introduced fresh water and spent her own money on hospital equipment and fruit and vegetables for the patients. Within six months, the mortality rate had fallen from 47.2% to 2.2%.

She invented a diagram which she called the 'coxcomb' – we now call it a pie chart – to show the effect her work had had. She sent this to MPs and civil servants who, she thought, would have been unlikely to read lengthy reports. After the war, she continued as a campaigner to improve nursing care and hospital administration, and made nursing a respectable profession for women. Her famous dictum, which now seems obvious but then was not, was that 'The very first requirement in a hospital is that it should do the sick no harm.' She died a spinster in 1910 at the age of 90, having published more than 200 books, reports and pamphlets.

OMEGA 3

IN 1972, a British professor made a startling prediction about the impact junk food would have on the health of billions of people in the developed world. Nutritional expert **Michael Crawford**, director of Brain Chemistry and Human Nutrition at London Metropolitan University, discovered that the brain needs essential fats – particularly Omega 3 'fatty acids' – for growth and development. He forecast that industrial farming (where meat- and milk-producing cattle and sheep are fed cereal, rather than grass) and a human diet featuring lower consumption of fish and seafood and more processed food, which is high in saturated fat, would lead to an increase in mental health problems.

Current NHS figures seem to bear this out: there are 250,000 admissions to psychiatric wards every year and increasing numbers of people are being diagnosed with conditions including Alzheimer's disease and depression (though there may be other factors in play). Building on Professor Crawford's research into brain starvation, the link between diet and anti-social behaviour was examined during a recent study into violence in prison. Researchers took 231 inmate volunteers at a prison in Aylesbury, Buckinghamshire, and gave half of them a placebo. The others were fed multivitamins, minerals and essential fatty acids. The number of violent offences committed in jail by those on the supplements fell by 37%. There was no change in the behaviour of those on the placebo and the violence levels returned to previous levels when the trial finished.

OPTICS

THOMAS Young (1773-1829) taught himself to read aged two, knew Latin at six and was fluent in 13 languages as a teenager. Yet it is his work in the field of optics which earns him a place as a medical great.

The oldest of ten children and the son of a Somerset banker, Young set up a medical practice in London, in 1799. While still a student, he discovered that the lens of the eye changes shape to focus on objects at differing distances. In 1801, he showed that astigmatism, which causes blurred vision, is due to the cornea being misshapen. He also came up with a theory of how the receptors in the eye perceive colours. He correctly proposed that the eye has three different colour receptors – red, green and blue – having worked out that the retina could not possibly have receptors for each individual colour. Other colours – such as pink or purple – are seen when the eye's 'cone' cells are stimulated in different combinations. White is produced by the combination of the three main colours, and black results from the absence of stimulation.

ORTHOPAEDICS

A TINY, chain-smoking Welshman who wore a long black frock coat and a sailor's cap pulled down over a dodgy eye holds the unlikely distinction of being the 'father of modern orthopaedics'.

Hugh Owen Thomas (1834-1891), was born in Anglesey, the son of a long line of bone-setters. He studied medicine in Edinburgh, set up practice in Liverpool's docklands and became an incredibly hard working doctor. He took just three days off a year, to visit his mother's grave in Anglesey, and treated 80 patients, mainly shipyard workers and their families, each day. His rounds started with home visits at 5am. He travelled in a scarlet two-seater carriage, pulled by two black horses, with his wife, Elizabeth, by his side.

After breakfast he would see patients in his surgery. Following lunch he would operate and at 8pm he would do his last house calls. After that he would work on new splint designs in his workshop or write in his study. Thomas, an accomplished flute player who stood just 5ft tall, was well ahead of his time, and **revolutionised the treatment of fractures**. He thought enforced rest and immobilisation were the best treatments for a broken limb, and employed a blacksmith and saddler to make splints. The 'Thomas Splint' – which extends from a ring at the hip to beyond the foot, and allows traction for a broken leg – is still in use today.

In 1870, he set up a free Sunday clinic for the poor, and was besieged by patients arriving in wheelbarrows and handcarts. On his death, of pneumonia in 1891, the streets were jammed with Liverpool's poor following the hearse to his funeral. His nephew, Brigadier Robert Jones, ensured his uncle's work was not forgotten. During WWI he was responsible for the evacuation and treatment of soldiers injured in the trenches. The death rate for men with a broken leg – especially where the bone broke through the skin – was 80%. Jones introduced his uncle's Thomas Splint, cutting the mortality rate to 20%.

PENCILS (AND ERASERS)

THE 'lead' in your pencil is actually a form of carbon called graphite, and it was first used to write with in Britain in the 1500s, after huge deposits of the mineral were found in the Borrowdale region of Cumberland. A violent storm in about 1560 uprooted many trees and monks discovered an unusual black material in the roots. They started using it to mark their sheep, calling it 'black lead', 'wad' or 'plumbago'. In fact, it was **the finest and purest graphite ever found anywhere in the world**.

Pencil-making started as a cottage industry in the region, with graphite being cut into sticks and wrapped in string or sheepskin. Cumberland graphite was much-prized by artists and writers – the Michelangelo School of Art in Italy used it – and it was so valuable that, in the 1700s, theft of wad was punished by a whipping and a year's hard labour, or seven years' transportation to Australia. The first pencil-making factory in the UK was set up in Keswick in 1832.

Of course, once you invent graphite pencils, you'll soon need an eraser. Draughtsmen used bread to rub out their mistakes, but in 1770 scientific instrument maker **Edward Nairne** found that rubber worked better. Born in Sandwich, Kent, in 1726, Nairne began selling 'rubbers' at his shop in Cornhill, London, at a costly three shillings per half inch cube. (Nairne also made microscopes, compasses and invented the **first successful marine barometer** to measure atmospheric pressure on board ships. It was taken on Captain Cook's second voyage to the south Pacific in 1772.)

PENICILLIN

PENICILLIN is one of **the greatest medical breakthroughs the world has seen**. But a British doctor's discovery of the power of this 'wonder drug' – so-called because of its ability to save lives which would otherwise certainly have been lost – came about by chance.

In 1928, **Alexander Fleming** (1881-1955) was looking at a culture of *Staphylococcus aureus* (a bacterium which commonly appears on the skin and in the noses of healthy people but can cause death from septicaemia if it enters the bloodstream) which he had been growing in a lab in St Mary's Hospital Medical School in London. He had been away for a fortnight with his family, and during his absence a rogue and unidentified airborne spore had landed in the dish and contaminated his experiment. As he peered closer, though, he saw that the stray spore was surrounded by a bacteria-free circle.

Whatever it was, it seemed to have inhibited the growth of the *Staphylococcus*. Excited, Scottish farmer's son Fleming carried out further experiments, and determined that the substance was indeed killing the bacteria. He isolated it – it was a mould called *penicillium notatum* – and named the active substance 'penicillin'.

However, it was not until the outbreak of WWII that penicillin was mass-manufactured and used widely. Two scientists, Howard Florey, an Australian, and Ernst Chain, a German refugee, working at Oxford University, did further testing of penicillin and turned it into a powder that maintained its antibacterial power for longer. The **world's first use of penicillin** was in treating a 43-year-old police constable called Albert Alexander, who was close to death after being scratched by a rose thorn. The infection had spread to his shoulder and lungs, he had lost an eye and had swellings and abscesses to his face. He received his first injection in Radcliffe Royal Infirmary, Oxford, in February 1941. Treatment continued for five days, and he made a rapid improvement. Sadly, supplies then ran out and he quickly regressed and died of blood poisoning. Further clinical trials were a huge success and it was realised that the new drug was needed immediately for the troops on the front line. Mass production started and penicillin saved many soldiers' lives. Within months, it was being used to treat bacterial infections in wounds as well as diphtheria, gangrene, pneumonia, syphilis and tuberculosis – all of which had often proved fatal.

In 1945 Fleming, Florey and Chain shared a Nobel Prize. Fleming died of a heart attack ten years later and his ashes were interred in St Paul's Cathedral.

PHILANTHROPISTS

THE Rowntree's sweets empire was started in York in 1862 by Henry Rowntree. But the driving force was his brother Joseph who turned the little local confectioner into a world giant, and also made his name as one of the first philanthropists of the new business age. When

Joseph joined in 1869, Rowntree's employed just 30 people. By the end of the 19th century it employed 4,000.

Joseph Rowntree (1836-1925) was a smart, forward-thinking boss. A Quaker, he set up an early occupational pension scheme, gave workers a week's paid holiday, appointed a company doctor and dentist, built a public swimming pool, opened schools and a gymnasium, provided dining facilities for thousands of employees, established one of the first widows' benefits funds and set up an unemployment scheme. In 1921, he opened 20-acre Rowntree Park in York as a memorial to company employees who had died in WWI. Rowntree trusts were set up, and they continue today, managing affordable housing and care homes for the disabled and elderly, and funding research into the causes of poverty and poor housing. Their sweets aren't bad, either. Rowntree's Fruit Pastilles hit the shops in 1881 and Fruit Gums were added in 1893. Black Magic chocolates came along in 1933, Aeros and Chocolate Crisp – now known as Kit Kat, and Britain's best-selling chocolate bar – were first sold in 1935 and Quality Street (Saddam Hussein's favourite chocs) and Smarties a couple of years later.

The Birmingham-based Cadburys were another philanthropic chocolate-making Quaker family. Tea dealer and coffee maker **John Cadbury** (1801-1889) started his cocoa powder business in 1831 and by 1853 he was appointed cocoa manufacturer to Queen Victoria. He devoted himself to the temperance movement and campaigned to replace 'climbing boys' with machines to sweep chimneys. Keen to improve working conditions in the late 1800s, Cadbury moved its operations to (then) rural Bournville, four miles out of Birmingham. They provided staff with housing, education, pension schemes and medical facilities. This ensured a healthy and dedicated work force and Cadbury's efficiency of operations improved. The company produced **the world's first chocolate Easter eggs** in 1875, Dairy Milk – called 'Dairy Maid' until six weeks before it was launched – was introduced in 1905, and the original dark chocolate, Bournville, arrived in 1908.

PHOTOGRAPHY

THE son of an English potter took **the world's first photographs**.

Thomas Wedgwood, born into the Staffordshire pottery dynasty in 1771, found that if he painted silver nitrate onto paper, white leather or glass, and then projected an image onto the surface, the silhouette of the object was left behind. Unfortunately, the pictures disappeared in any light stronger than a candle. He might have found a way of 'fixing' his photos, but his health had never been good and he died aged 34 before he had chance.

In fact, Frenchman Joseph Nicéphore Niépce took the first permanent picture in 1827; he used bitumen mixed with lavender oil painted onto a sheet of pewter as his medium, but his photograph, taken using a pinhole *camera obscura* and showing a building and a tree, took eight hours of light exposure.

The eminent Dorset mathematician and MP **Henry Talbot** created **the world's first 'negatives'** in 1834, using paper soaked in silver chloride and fixed with a salt solution. The earliest paper negative in existence, produced in August 1835, shows the latticed lead lights of a window at his home, Lacock Abbey (now owned by the National Trust and open to visitors) in Wiltshire. Positive images were made by contact printing onto another sheet of paper. These still faded over time, but **John Herschel**, son of British astronomer William (see *Uranus*), helped him with the problem of fixing the image permanently in 1840, using sodium thiosulphate to wash out the unchanged silver salts in the picture. Herschel also coined the words 'photograph', 'snap-shot' and 'negative'.

Talbot's photos were not very detailed and contained all the imperfections of the paper. If an image could be fixed to a glass plate, this would solve that problem. But how to make the silver nitrate stick to the glass? In 1848, another Frenchman started coating glass plate with eggwhite and potassium iodide. This allowed for very detailed photographs, but was extremely slow – portraits were impossible, as the subject could not remain still for long enough, and it was only suitable for photographing buildings and landscapes.

In 1851, a butcher's son from Bishop's Stortford, **Frederick Scott Archer**, solved this problem by inventing the 'Collodion' process. Collodion, cellulose nitrate dissolved in ether and alcohol, had been used as a wound dressing (it dries and forms a 'second skin'). But when applied to photographic plates in conjunction with silver nitrate it was much faster than previous methods, reducing exposure times to a second or two. It allowed for **the world's first proper photographic portraits** to be taken. Archer published his findings in 1854, but did not patent the technique and thus made nothing from it.

Although very effective, Collodion was inconvenient and messy – it required that coating, exposure and development be done whilst the solution was still wet, and that substantial quantities of chemicals and equipment be lugged around by the photographer. A British doctor, **Richard Leach Maddox**, invented a dry substitute using gelatin in 1871 – this led to the introduction of mass-produced plates and marked the beginning of the modern era of photography, because it allowed the development of relatively small, portable cameras. Unfortunately, Maddox, too, neglected to patent his discovery and he died in poverty in 1902.

Scottish physicist **James Clerk Maxwell** (see *Electromagnetism* and *Microwave Ovens*) produced **the world's first colour photograph** in 1861, though not as we know it. At a meeting of the Edinburgh Photographic Society, he projected onto a screen three images of a tartan ribbon, each taken through different colour filters. When overlapped, the colours combined to form one 'colour' image. (The image is displayed at a museum in Edinburgh in the house where he lived.)

The great developments in photography thereafter came in the USA, when George Eastman developed mass market film cameras for the ordinary person. But mention should be made of **Edward Muggeridge**. Born in Kingston-upon-Thames in 1830, he travelled to America to seek his fortune as a young man. Having changed his name for obscure reasons to Eadweard Muybridge, he began importing English books and took up photography. In 1872, a wealthy race horse owner called Leland Stanford asked Muybridge to use his photographic skills to settle an argument: whether a race horse ever had all four of its feet off the ground at once as it ran.

He was about to start work on this burning question when the small matter of a murder trial intervened. His wife – Flora Shallcross Stone, a divorcée half his age – was having an affair with one Major Harry Larkyns, an English drama critic. Muybridge intercepted a letter sent by the Major to Flora. Consumed with jealousy, he paid a visit to his rival. Flourishing the letter, he said, 'Good evening, Major. My name is Muybridge, and here is the answer to the letter you sent my wife.' With that, he pulled out a gun and shot the unfortunate adulterer dead. He was acquitted of murder – the court decided the killing was a justifiable homicide – and was able to resume his horse-related investigations.

In 1878, he arranged twelve cameras in a line, 21 inches apart and set to take photographs at intervals to cover a 20ft stretch of racetrack as a horse came by. In doing so, he proved that there *is* a moment when all four feet are off the ground as a horse gallops – and while the usefulness of this was limited, it was the first time photography was used to solve practical questions which the naked eye could not answer.

PILL COATING

BITTER pills became easier to swallow in the 1800s, thanks to an innovative English chemist. **Arthur Hawker Cox**, a cabinet-maker's son who was born in London in 1813, was working as a chemist's apprentice for his uncle in Northampton when he came up with the idea of coating tablets to disguise their horrible taste. After years of experimenting, he was issued a patent in 1854 for a tasteless shell for medicinal tablets. Cox, who moved to Brighton after marrying Mary Anne Strudwick in 1837, became a politician and was Mayor of Brighton three times.

SAMUEL PEPYS

WHEN **Samuel Pepys** began writing a diary on January 1, 1660, he could scarcely have imagined that it would turn out to be one of **the world's most important historical documents**, still widely read almost 350 years later. Dramatic events including the plague of 1665 and The Great Fire of London the following year were all vividly recorded in his neat shorthand, along with more mundane events, his thoughts and fears and everyday happenings in London.

Born in the capital on February 23, 1633, the fifth of eleven children, Pepys was the son of a Cambridgeshire tailor. He became an MP and Chief Secretary of the Admiralty – driving the increasing professionalism of the Navy – but we remember him today for his diaries. They ran to 1,250,000 words over nine years, and contributed greatly to our knowledge of life during that period. Unfortunately, he stopped writing in the spring of 1669, aged just 36, because he feared that he was going to lose his sight through eye strain. He died in 1703 and bequeathed his library of 3,000 books and manuscripts to Magdalene College, Cambridge, where he had studied, on the understanding that its contents would remain unaltered and intact. The collection is still there, in the bookcases he had made for it.

PLASTICS

PLASTIC is used everywhere from the cockpit of the Space Shuttle to the toothbrush in your bathroom – and **the world's first plastic** was invented in 1862 by a metallurgist from Birmingham. His name was **Alexander Parkes** and, modestly, he named it '*Parkesine*'.

It was made by reacting cellulose (extracted from wood pulp) with nitric acid. The resulting material – chemically, cellulose nitrate – was used to make buttons, billiard balls and combs which Parkes displayed at the London International Exhibition that year. To the Victorians, this was a revelation: a lightweight, man-made substance, transparent or opaque, easily coloured, and capable of being moulded ('plastic' comes from the Greek word *plastikos* meaning 'can be moulded'), carved or spun into fibre.

It was not without its drawbacks, however. The billiard balls made an alarming bang when they collided at speed (a Colorado bar owner who had bought some claimed that the sudden noise resulted in every customer drawing his gun) and it was extremely flammable. It was joked at the time that the perfect Christmas present for one's mother-in-law was a Parkesine dress and a packet of cigarettes.

Parkes was an enthusiastic inventor, taking out no less than 80 patents during his life. He carried his enthusiasm over into other areas of endeavour, too, fathering 20 children by two wives.

British chemists **John Whinfield** and **James Dickson** made **the first polyester** (see *Synthetic Fibres*) 80 years later and many other plastics have followed.

The most amazing of them – potentially – is 'Starlite', created by accident 20 years ago by a former ladies' hairdresser from Hartlepool called **Maurice Ward**. Ward, an amateur chemist who had set up a small plastics business, was playing around with various chemicals and somehow produced **the world's most heat-resistant material**. Actually, 'heat-resistant' doesn't really tell the full story: Starlite can withstand temperatures three times that at which diamonds burn. He has been in discussions with NASA (his plastic has an energy absorption rating 2,470 times better than the heat shield tiles used on the Space Shuttle), aircraft manufacturers and governments. The British Atomic Weapons Establishment subjected it to simulated nuclear explosions at 10,000°C, twice, and couldn't destroy it – whereas pure carbon, which has the highest vaporisation point of all elements, melts at 3,500°C.

If this all sounds too good – and strange – to be true, you can see for yourself: visit YouTube and search for 'Starlite and *Tomorrow's World*'. You'll find a video from the respected BBC TV science programme from 1990, in which a presenter tries to cook an egg coated with a thin layer of Starlite using an oxyacetylene torch. After five minutes of burning at 2,500°C, the egg is still cool to the touch, and when it is cracked open it is still completely raw.

Frustratingly, given its astonishing, almost limitless potential, Ward has so far been unable to agree terms for the commercial use of the plastic. His lawyer, Mishcon de Reya partner Toby Greenbury, told the *Sunday Telegraph*: 'Maurice is a one-man band. He's an inventor, and he has an unusual way of looking at things. It has proved to be very difficult to deal with large companies. I would really like to see this commercialised in Maurice's lifetime. It's difficult to think of another invention that is bigger in its implications.'

POP MUSIC

THE greatest classical composers – men like Beethoven, Mozart and Bach – have all been continental Europeans (though Edward Elgar and Benjamin Britten have made serious contributions to the art, and Handel was an adopted Brit).

But we've definitely produced **the world's greatest pop music**.

A search for 'French pop greats' on Google produces precisely three hits, and some discussion of Serge Gainsbourg and Sacha Distel.

A 'German pop greats' search brings a princely *nine* hits (after all, they produced the Spinal Tap-like Scorpions, electro-bores Kraftwerk and something about red balloons).

In fact, only the USA runs us close, and we changed *their* popular music with the 'British Invasion' of 1964. That was led by The Beatles, **the most famous group of all time**. To date, the Fab Four have sold a billion records – almost certainly more than every artist in the rest of Europe put together. That's without mentioning The Rolling Stones, David Bowie, The Clash, The Kinks, Eric Clapton, Led Zeppelin, The Who, Pink Floyd, Oasis, Radiohead, Blur, Coldplay, The Stereophonics...

THE PORTABLE DEFIBRILLATOR

COUNTLESS heart attack victims around the world owe their lives each year to a Briton who almost died in a Japanese prison camp during WWII.

Dr James Pantridge (1916-2004) was used as slave labour on the Burmese Railway project immortalised in the film *The Bridge Over The River Kwai*. Emaciated and suffering from cardiac beriberi – a disease caused by a vitamin B1 deficiency which can lead to heart failure – he weighed just 80lbs (30kg) when he was liberated.

That appalling experience, and the scary cardiac death rate in the UK, gave him an interest in heart disease. The Swiss had first experimented with defibrillation – the process of electrically shocking a failing heart back into a healthy rhythm – in the late 1800s, and the American/Lithuanian Bernard Lown made advances in the 1950s. But by the 1960s, the few unwieldy machines which existed were in hospitals, and many heart attack victims died before they got to them. Pantridge, of Hillsborough in Northern Ireland, decided to find a way to take intensive care facilities to the patient. He created **the world's first portable defibrillator** in 1965. Weighing 70kg, and powered by two car batteries, it was installed in ambulances. Within three years, defibrillators weighed just 3kg. Pantridge's device had revolutionised emergency medicine and meant that the heart could be 'jump-started' anywhere.

A bachelor and keen salmon fisherman, who had won a Military Cross for saving lives under heavy bombing and shellfire in the war, Pantridge died at the Royal Victoria Hospital, Belfast on Boxing Day, 2004. The cause of death: heart attack.

PUBLIC PARK

IN the Victorian age, millions lived in cramped, back-to-back terraces, breathing in choking, smog-filled air. They desperately needed somewhere green and pleasant to escape the grimy streets – and **the world's first public park** was the answer.

Derby Arboretum was opened to the public on September 16, 1840. **Joseph Strutt**, a 75-year-old mill owner and mayor, gave the gardens to the people of the town. At the opening ceremony, he declared that the park 'shall be open to all classes of the public without payment, and subject only to such restrictions and regulations as may be become necessary for the observance of order and decorum'.

Birkenhead Park, in Liverpool, officially opened on April 5, 1847, claims to be **the world's first wholly-publicly-funded park**. Recently restored at a cost of £11.25 million, it was the inspiration for Manhattan's Central Park, whose designer, Frederick Law Olmstead, had visited Birkenhead.

PURPLE (AND PERFUME)

A GIFTED young chemist from London made the world a brighter and more fragrant place with two discoveries. Aged just 18, builder's son **William Perkins** (1838-1907) was trying to synthesise quinine – a treatment for malaria made from the bark of the cinchona tree, which was costly and hard to come by in its natural state. He didn't succeed but when he mixed coal-tar extracts with 'aniline' – a substance still widely used in making dyes, drugs, plastics and explosives – and alcohol, he found the resultant concoction was an intense purple colour. Until that happy accident, purple dye was so difficult and expensive to produce (it was made by boiling up molluscs) that it was reserved for royalty, peers and bishops. Perkins patented his product and set up a factory in north London with his father, George. It made him rich; Queen Victoria wore a silk gown coloured with his purple dye to the Royal Exhibition in 1862.

In 1867, when still a young man, he also revolutionised the perfume industry by creating **the world's first synthetic scent** – a man-made version of coumarin, a vanilla-like fragrance found naturally in American tonka beans. It was the start of the modern age in which perfume became affordable to the average person.

RADAR

RADAR, which allows operators to 'see' aircraft on screens at distances of many miles, is a British invention. **Robert Watson Watt** – the son of a Scottish carpenter, and a descendant of steam pioneer James – first demonstrated a working system in 1935.

Watt, who was born in Brechin, Angus, in 1892, excelled in applied mathematics and electrical engineering at university and spent the 1920s working on ways to use radio to detect thunderstorms. In the mid 1930s, he was asked by the War Ministry to turn his expertise to developing a 'death ray', because the Nazis had recently claimed *they* had such a weapon. He quickly scotched this idea as an impossible piece of propaganda, but mentioned that he thought he could build equipment to detect aeroplanes in flight. He was immediately asked for a demonstration, and at a secret test site near Daventry on February 26, 1935, he successfully showed that he could locate an RAF bomber flying nearby. (Radar works by transmitting radio waves or microwaves into the atmosphere; these hit the target aircraft and bounce back to a receiver.) In April, he was granted a patent for his invention, though the term 'radar', an acronym for RAdio Detection And Ranging, wasn't used until 1941.

By the outbreak of WWII, 19 radar stations were up and running and they were vital in saving Britain from invasion by Hitler's forces. Its early warning of incoming German air raids allowed the RAF to compensate for the numerical superiority of the Luftwaffe by directing our Hurricanes and Spitfires to intercept them. Had we been 'blind' we would very likely have been overwhelmed.

Watson Watt, a short, tubby and verbose Socialist, received £52,000 (£5 million today) from the Royal Commission on Awards to Inventors after the war. He died Sir Robert in 1973. Of course, as with most other inventors, he built on the work of others, here and abroad. As well as the German Christian Hülsmeyer and the Serb Nikola Tesla, many others contributed to its realisation, including John Logie Baird, Alan Blumlein, Geoffrey Dummer, Godfrey Hounsfield, Christopher Cockerell, John Randall and Harry Boot – all mentioned in this book for other things.

RADIO

A BRITON, Welshman **David Hughes**, was **the first person to transmit and receive electromagnetic radio waves**. His achievement, demonstrated in London to members of the Royal Society in 1879, represented the birth of radio. They were unimpressed, and the results were never published. In fact, although it was years before the Italian Guglielmo Marconi's more celebrated invention (in Britain) of the radio telegraph in 1897, Hughes' achievement is often glossed over even now.

However, it is not all he is known for. Having been born either in Merioneth, Wales, in 1829, or in London of Welsh parents in 1831 (the records are unclear), he emigrated to America. It was there, as a young man, that he invented a new form of type-printing telegraph which revolutionised the burgeoning communications industry. It was adopted by the American Telegraph Company in 1855 and Hughes returned to Britain with his invention in 1857. It was not well received, but the French government bought it in 1860 and installed it across France. Over the next decade it spread across Europe – including Britain – and was the basis for what became a substantial fortune.

A walrus-moustached *bon viveur* – he is described in the *Oxford National Dictionary of Biography* as 'a genial and charming man of simple tastes and often the life and soul of the party at the regular lunches he attended with his friends' – he also invented the **modern**

microphone. He refused to seek a patent for the device and instead generously gave it to the world: it allowed the development of telephones but also the Eurovision Song Contest, so there were pros and cons. He died from flu in London in 1900, and left a fortune of £469,491 11s 9d – £200 million today. Having no children, he made sure his widow would be taken care of and donated the remainder of his estate – well over £400,000 – to hospitals and scientific establishments in the capital and across the continent. Incidentally, **the world's first radio station** was set up by Marconi in 1897 at Alum Bay, Isle of Wight. The landlord who owned the building responded to its success by putting up the rent, and the station closed down in 1900.

ROLLS-ROYCE

THE phrase 'Rolls-Royce' is synonymous with quality, thanks to **the world's most famous luxury car brand**.

Rolls-Royce Limited was founded by two Britons from very different backgrounds. **Charles Rolls** (1877-1910) was the flamboyant and Eton-educated son of a wealthy and aristocratic Conservative MP, while **Henry Royce** (1863-1933) was born the son of a poor flour miller from Alwalton, near Peterborough.

Henry was only nine years old when his father died – leaving less than £20 – and he struggled through his early life with limited education and part-time jobs. But an aunt paid for an apprenticeship at the Great Northern Railway's locomotive works, he showed a great aptitude for electrical and mechanical engineering and, by the turn of the 20th century, he was a successful Manchester businessman, building cranes and – by 1904 – making his first motor car.

Down in London, meanwhile, Charles Rolls had developed an obsession with cars. In 1903, he opened one of Britain's first car showrooms in the West End of London. Unfortunately, all the motors were foreign, and he was desperate for a high-class British vehicle to sell. The following year, Rolls saw pictures of Royce's first car and travelled to his Manchester workshop. They joined forces, with Royce as the engineer and Rolls the marketing man, and soon opened a factory to build the new 'Rolls-Royces' in Derby. Their first was the '10hp' – a 10 horsepower, open-topped machine capable of 39 mph, of which 16 were made. (The list price was £395; in 2004, a roadworthy example was sold for £3.5 million.) In 1907, their '40/50' was launched. Built to Henry Royce's exacting standards of construction, it was astonishingly reliable (many of the 7,874 made over the following 20 years are still running) and it was named 'The best car in the world' by the (then) highly-prestigious *Autocar* magazine. Later models of this vehicle were named the Silver Ghost because of their quietness, and it was this vehicle which 'made' the Rolls-Royce brand.

For decades, the 'Roller' – from the Silver Ghost to the Phantoms, Silver Wraiths, Clouds and Shadows, and the Corniche – was the preferred vehicle of kings, queens and heads of state around the world. John Lennon was among the first celebrities to buy one. He owned a Phantom V, adding a radio telephone with the number Weybridge 46676, a double-bed for the rear seat, a sound system with a loud hailer, a television and a portable fridge. He also had the bodywork painted in psychedelic colours and was once attacked with an umbrella by an elderly woman in London for 'ruining a Rolls-Royce'. After Lennon's death, it was sold at auction in 1985 for what would be, today, more than £3 million.

Unfortunately, Charles Rolls did not live long enough to witness much of the great success of the brand he had helped to create. He was just 32 when, competing in a flying tournament in Bournemouth in July 1910, the tail of his plane buckled and he crashed and died. Henry Royce had better luck. From his humble beginnings, he made a fortune and was knighted, though his health suffered after the early death of his friend and partner, and he could be a difficult man to like. When Rolls-Royce bought out their rivals Bentley Motors in 1931, for instance, Royce made the brilliant founder and designer Walter Bentley work as a sales assistant in his London showroom.

Rolls-Royce also made marine and aircraft engines – the legendary 'Merlin' powered the Spitfire, and for a time **the world land, water and air speed records** were all held by machines containing the company's engines. Unfortunately, its work on the brilliant but costly RB211 turbofan bankrupted the company in 1971. It was nationalised, and the car and aero engine divisions were split. Later re-privatised, Rolls-Royce Motor Cars is now owned by Germany's BMW, while the British-owned Rolls-Royce plc is a world leader in aero engines and supplies Boeing, Airbus and military planes (as well as ships and subs).

RUGBY UNION

LEGEND has it that rugby started at Rugby School in 1823 when a boy called **William Webb Ellis** picked up the ball and ran with it while playing football (which, as we've seen, was not a formally organised sport at that point). Whether this is true or not, the Rugby Football Union was created in 1871. Rugby Union is now the main sport in New Zealand, South Africa and numerous south Pacific islands, as well as being played enthusiastically in Australia, Canada, the USA, France, Argentina, Japan, India and many other countries around the world. Unusually, for a sport invented on these islands, one of the home nations was, for a time, the undisputed best team in the world. In 2003, under Clive Woodward and Martin Johnson, England beat everyone put in front of them (including New Zealand in New Zealand and Australia in Australia – twice) and won the Rugby World Cup.

Rugby League, the 13-man version of 15-man Rugby Union, originated in Yorkshire, Lancashire and Cumbria over a century ago. Working men from poor northern towns couldn't afford to play rugby without being paid, so sought money from their clubs. The southern-based (and very middle class) RFU insisted on its players staying amateur and unpaid. In 1895, the northerners left the RFU and later created a similar but recognisably different game which they successfully exported to Australia, New Zealand and elsewhere.

SAFARI PARKS

A BRITISH circus owner set up **the world's first safari park**, creating a storm of protest in the process. Inspired by a visit to Kenya, **Jimmy Chipperfield** decided to open a new type of zoo, where animals could have space to wander in large enclosures resembling their natural habitat. Chipperfield, who was born in a circus wagon in Wiltshire in 1912 and toured the world as a clown, bear-wrestler and lion-tamer, persuaded the 6th Marquess of Bath, Henry Thynn, that the grounds of Longleat, his ancestral family home in Wiltshire, would be the perfect location.

His plan was for a 100-acre reserve for 50 lions, extras from the film *Born Free*. People asked how big the cages would be. Chipperfield replied: 'It's the people who are going to be in the cages (their cars) and the lions who are going to be free.' It created huge controversy, with fears that the lions could pull visitors from their vehicles and eat them, or escape and rampage across the English countryside. *The Times* campaigned against Longleat, declaring solemnly: 'No amount of soothing assurance can persuade sensible people that a quite gratuitous and unnecessary risk to life is not contemplated.' When the gates opened for the first time, on Easter weekend 1966, the traffic queues stretched for miles. No-one was eaten.

The costs of the project were repaid within five months and Chipperfield's foresight changed the way animals were kept in captivity, improved breeding programmes, and paved the way for more drive-through safari parks to open around the world.

SANDPAPER

JOHN Oakey, a piano maker's apprentice from Walworth, London, who was born in 1813, started mass-producing sandpaper in 1833. He also made 'Oakey's Knife Polish' and in 1858 used **the world's first celebrity endorsement** when he stuck the Duke of Wellington's image on tins of the stuff. The fact that the Duke had died six years earlier did not seem to affect sales.

SAS (AND SBS)

THE world's leading special forces are British.

The SAS – Special Air Service – was started in 1941 as an elite regiment which would operate behind enemy lines. Made up of (mostly) British Army soldiers who pass a gruelling selection test to enter, it has become the model for the special forces of most other countries, including the USA's 'Delta Force'. (The less famous Special Boat Service, or SBS, is the Royal Marines equivalent of the SAS.)

The unit was set up by **Lieutenant David Stirling** (1915-1990). The forceful, 6ft 6in tall son of a Brigadier-General from Kier, Stirlingshire, he wanted to launch daring, hit-and-run attacks on Nazi airfields, supply depots and forces deep in German-held territory during the desert fighting in North Africa. He and his men were incredibly brave and resourceful; on occasion, they drove straight through enemy camps disguised as Germans, which would have resulted in their immediate execution had they been discovered, and they destroyed hundreds of planes, vehicles and stores.

Since its early days, the SAS has fought in every conflict involving British forces, as well as evolving into **the world's leading counter-terrorist unit**. It played a key part in the military defeat of the PIRA in Northern Ireland, and is currently heavily involved in the Middle East.

It is notoriously secretive but some light was shed on its activities in May 1980, during the Iranian Embassy siege. A group of dissidents seeking autonomy for a breakaway region of Iran had seized the embassy and taken hostages. When they killed one and threw his body outside, Prime Minister Margaret Thatcher ordered the SAS to storm the building. A worldwide TV audience watched as 20 black-clad men abseiled onto the roof from helicopters and used explosive charges to blow their way into the building. All but one of the 26 hostages were released unharmed and five of the six terrorists were killed. Mrs Thatcher's husband Dennis famously asked later that day,

'Why did you let one of the bastards live?' In a sign of how times
have changed, the surviving 'bastard' – Fowzi Nejad – was released
from prison in 2008 and allowed to stay here with a new identity and
a secret address funded by the British taxpayer.

SCOUTING

SCOUTING was started in 1907 by the British Army officer **Robert
Baden-Powell** (1857-1941). He held a camp for boys on Brownsea
Island in Dorset and the following year published *Scouting For Boys*,
adapted from a military guide he had written. The movement – which
soon encompassed cub scouts, girl guides and brownies – emphasised
practical outdoor activities like camping, woodcraft and hiking. The
distinctive uniform with neckerchief and 'woggle' was introduced to
hide differences in wealth or social standing and the 'merit' badges
demonstrated achievement in a variety of fields. Today there are close
to 40 million scouts and guides in more than 200 countries.

THE SEISMOGRAPH

A DEVASTATING earthquake which rocked Japan led an English
geologist to invent the seismograph. **John 'Earthquake' Milne**,
who was born in Liverpool in 1850, also set up the world's first
laboratory to study the phenomenon. Milne, a wool merchant's
son, moved to Tokyo in 1875 to work as a professor of geology and
mining. In February 1880, the city of Yokohama was badly damaged
by an earthquake. The disaster led Milne to embark on years of study
into tremors, and in 1896 he unveiled his seismograph (created with
the help of his fellow British scientists **James Alfred Ewing** and
Thomas Gray). It measured vibrations in the earth and recorded
them on photographic paper.

THE SEWING MACHINE

A LONDON cabinet maker named **Thomas Saint** invented
the world's first sewing machine in 1790, though he is not

always credited with the invention because his patent was misfiled and lost for nearly a century. He had patented several items at the same time, including a method of clog-making and a glue, and they had all been placed in the same patent folder under the name 'Glue'.

His hand-operated machine was designed to sew leather using a chain stitch. William Newton Wilson – a sewing-machine manufacturer who found the forgotten drawings at the patent office – made a replica using Saint's original sketches in 1873. The device looked more like a printing press and did not work until Wilson made minor modifications, but it is quite possible that Saint had deliberately filed an imperfect patent to prevent industrial espionage. Patents are obviously valuable, and in previous times copies were sometimes taken and sold to the highest bidder.

WILLIAM SHAKESPEARE

NEXT time you feel 'lonely', 'gloomy' or 'suspicious', you can thank **the world's greatest-ever writer** for adding those descriptive words (and around 1,700 others) to the English language. Equally, if you call someone a 'green-eyed monster', say they have a 'heart of gold' or complain that you 'have not slept a wink', you are citing **William Shakespeare**. According to *The Literary Encyclopedia*, he is the 'world's most quoted poet and dramatist'.

Shakespeare's most famous works have been translated into 80 languages (including *Star Trek*'s Klingon) and his popularity continues almost 400 years after his death – not least because of his uncanny understanding of human nature. 'After God,' said Pushkin, Russia's pre-eminent poet, 'Shakespeare is the greatest creator of living beings.'

His date of birth is believed to be St George's Day – April 23 – 1564. His parents, John, a glove maker, and Mary, lived in Stratford-Upon-Avon, Warwickshire, and William attended the local grammar school. He married Anne Hathaway when he was only 18 and she was 26, and she gave birth to their first child, Susanna, six months later. Their twins Hamnet (who died, possibly of the plague, aged 11) and Judith were born in 1585.

The success of works like *Macbeth*, *Hamlet* and *King Lear* enabled him to buy the second largest house in Stratford, New Place, in 1597. He is believed to have paid £120 for it, and he bequeathed it to his daughter Susanna after his death on April 23, 1616, aged 52. He also left £10 to the poor of Stratford and famously specified that his wife should have his second-best bed.

THE SHARKSKIN SWIM SUIT

THE 'Fastskin' swimsuit made a huge splash when it was launched at the 2000 Sydney Olympics. Over 80% of the swimming medal winners in Australia wore the new body suit, created by British designer **Fiona Fairhurst**. There were even calls to ban it because it seemed to give its wearers an unfair advantage over everyone else; this was rejected, as long as it was made available to all. Since then, 36 world records have been broken by swimmers wearing Fastskin, and the very best have shaved 3% off their own fastest times.

Fairhurst, a biomimetician who studied textile technology at university in Huddersfield and was a competitive swimmer until she was 16, wanted to decrease the water drag which slows swimmers down. The breakthrough came during a visit to the Natural History Museum, thanks to Senior Fish Curator Oliver Crimmen and the pickled shark he showed her. Fairhurst examined the shark's skin under a microscope and saw that it contained hundreds of tiny ridges. These 'denticles' reduce friction and – together with a streamlined shape – allow sharks to be hugely dynamic underwater, despite their size and weight. Working with swimwear manufacturer Speedo, she came up with a water repellent fabric which mimicked that effect.

SHORTHAND

THE most widely used shorthand system in the English-speaking world was devised by a Briton, **Sir Isaac Pitman**, (1813-1897) from Trowbridge, Wiltshire. He came up with the system of dashes, curves and dots to represent the phonetic alphabet in 1837. In a letter to his brother Joseph, Pitman said he had 'contrived a very good alphabet... (It is) a scientific work, which, I think, will stand the test of years.' He was right – it enabled generations of journalists and secretaries to keep up with the spoken word.

SIGN LANGUAGE

THE inventor of the first sign language alphabet was a Scottish school teacher. **George Dalgarno** (1616-1687), who had nine children, was born in Aberdeen but taught for about 30 years at a Grammar school in Oxford. In 1680, he published *The Deaf and Dumb Man's Tutor* – a teaching method for educating deaf children. He also developed a universal language, but it failed to catch on.

SLAVERY (ABOLITION OF)

WILLIAM Wilberforce had a big nose, came from Hull and stood a mere five foot four inches tall. None of this stopped him from cutting a swathe through the ladies of London in the 1770s. A drinker, gambler and general carouser, Wilberforce was a wealthy man after receiving large inheritances, and he became an MP at the age of 21 by the traditional method of buying his votes (he paid merchants ten guineas and ordinary people two). He was an all-round unserious figure until undergoing a conversion to evangelical Christianity at the age of around 25. After that, Wilberforce (1759-1833) became convinced of the need for social reform and to bring an end to slavery. In doing so, he became **the world's most famous anti-slavery campaigner**, with his 1833 Bill for the Abolition of Slavery which outlawed the practice throughout the British Empire. He died just three days after the bill was passed.

He was also a driving force behind the creation of the Society for the Prevention of Cruelty to Animals (now the **RSPCA**) which is **the world's oldest animal welfare organisation**.

SMOKING AND LUNG CANCER

SIR Richard Doll and **Sir Austin Bradford Hill** made medical history in 1950 when they proved that **smoking causes lung cancer**.

At that time roughly 80% of British adults smoked, and there had been an alarming increase in the numbers dying from the disease. In 1947, Bradford Hill, a quiet, unassuming medical statistician and Doll, an epidemiologist, were asked by the Medical Research Council to find out why this was. Initially, as Britain was in the midst of a huge road-building programme, they thought fumes from car engines or tarmac-laying might be to blame. But then they discovered that all but two of 700 lung cancer patients interviewed in London hospitals were smokers. Doll, from Hampton Hill, Middlesex, promptly gave up himself and the government accepted the link in 1954 (the health minister, Iain Macleod, made the announcement while chain-smoking). But many medical experts, lots of whom liked a fag themselves, were sceptical. So Hampstead-born Bradford Hill (1897-1991) devised a new study – this time quizzing 40,000 doctors about their smoking habits. The number subsequently dying from lung cancer was recorded. That study continued for 50 years and proved beyond doubt that smoking caused the disease. Now only around a quarter of Britons regularly smoke.

Regarded as one of the most eminent scientists of his generation, Doll, whose father was a doctor and mother a concert pianist, also established that tobacco caused other cancers, discovered that alcohol increased the risk of breast cancer and found that aspirin protected against heart disease. He believed that the elderly should live life to the full, and proved the point by flying a glider and riding a camel in the Arabian Desert in the months before his death, aged 92, in 2005.

SONAR

A SHY mathematician from Newcastle-upon-Tyne devised **the world's first Sonar system** in 1912.

Active Sonar (an acronym for SOund Navigation And Ranging) works in a similar way to radar, by sending out sound waves through water and collecting them when they bounce back off a submerged object. In the aftermath of the *Titanic* disaster, **Lewis Fry Richardson** (see *Weather Forecasting*) suggested this might be a way of locating underwater hazards like icebergs. It was soon picked up by the military – the 'ping' sound effect from movies like *The Hunt For Red October* represents attempts by an enemy vessel to locate a sub using sonar.

Other countries soon copied the idea, but British sonar systems – called ASDIC, from Anti-Submarine Division Investigation Committee – were the best in the world throughout the First and Second World Wars (in fact, we gave our technology to the Americans in 1940).

THE SPARE WHEEL

RESOURCEFUL Welsh brothers came up with an invention that motorists now take for granted – **the world's first spare wheel**. Early drivers were often left stranded if they got a puncture because early cars did not come with a spare. **Thomas** and **Walter Davies**, from Spittal, Pembrokeshire, saw a gap in the market and made themselves a fortune.

In 1895, when Walter was aged 18 and Thomas 21, they moved to Llanelli and set up an ironmonger's shop. Soon they were also making bicycles and by 1902 they also had a car hire business. The idea for the 'Stepney Spare Wheel' – named after the street where their business was based – came to them after a judge had a puncture in the hire car as he was heading to court. It was unveiled in 1904 and was an overnight success. At their peak they sold 2,000 a month all over Europe and the United States.

SPECTACLES

EARLY specs had to be held on to the face or were tied on with silk ribbons. Londoner **Edward Scarlett**, optician to King George II, made a breakthrough in 1730 when he made **the world's first modern glasses** featuring rigid side arms which balanced on the tops of the ears.

A medical instrument maker from London, **James Ayscough**, made the first *hinged* spectacles, advertising them for sale in 1752. They were also the first glasses to have tinted lenses. Ayscough believed that blue- or green-tinted lenses were gentler on the eyes. Thus, **the world's first sunglasses** also came from Britain – although admittedly Ayscough's shades were not specifically designed to shield the eyes from the Sun.

SPEED

A BRITON holds the world land speed record – but then it's a record we've almost always held. **Andy Green**, a former RAF fighter pilot born in Atherstone, Warwickshire, became **the first man to break the sound barrier on land** in a supersonic, jet-powered car called Thrust SSC (SuperSonic Car), reaching 714.144 mph at Black Rock Desert, Nevada, in September 1997. Green is now planning to smash his own record in 'Bloodhound SSC', which has been designed at Swansea University; when built (funding is still being sought) it will accelerate from nought to 60mph in one second and touch 1,050mph – fast enough to overtake a bullet fired from a revolver.

It's all a far cry from Richard Trevithick's 'Puffing Devil' of 1801, when – for a time – the world's fastest vehicle could be overtaken by a half-decent jogger. The first proper record was set by the Frenchman Count Gaston de Chasseloup-Laubat in 1898, who travelled at 39.24mph in an electric car. But it was soon in British hands.

In 1924, the bespectacled Hampstead-born racing driver **Ernest Eldridge** became the last man allowed to set such a record on a public highway, in a car called 'Mephistopheles'. Powered by a 21.7 litre six-cylinder WWI airship engine, the vehicle took up the width of the tree-lined road in Arpajon, France, and veered wildly from side to side as it roared along the asphalt. It had no front brakes, and neither Eldridge nor his mechanic John Ames wore a crash helmet. (Ames, clinging on for dear life, was employed to regulate the oxygen supply to the engine and maintain the fuel pressure using a hand pump.) At any time, a farm animal or random Frenchie could have strayed into the road, the tyres could have blown out or the engine exploded, but none of this happened and the pair set a record of 146.01mph without, astonishingly, killing anyone.

Eldridge did not hold the record for long. Two months later, another Englishman, **Sir Malcolm Campbell**, from Kent, took the first of ten land speed records he was to hold when he hit 146.1mph on Pendine Sands, a stretch of beach in Wales. **Henry Segrave**, an Eton-educated racing driver, smashed the 200mph hurdle in March 1927 at Daytona Beach, Florida, but Campbell went past 300mph within a few years. He passed on his competitive streak and love of racing to his son Donald, who became **the first and only man to hold both land and water speed records in the same year**.

Born in Horley, Surrey, in 1921, he hit 403.1mph in 'Bluebird' at Lake Eyre salt flats in Australia in July 1964. In December, he reached 276.33mph in his jet-powered boat – Bluebird K7 – on Lake Dumbleyung in Perth, Western Australia.

Donald Campbell was killed on January 4, 1967, while trying to break his own water record and go past 300mph on Lake Coniston, Cumbria. In order for the record to be recognised, he had to complete two 1km journeys across the lake. On the second leg, hitting 320mph, the boat flipped over and disintegrated. He died instantly – and cruelly, because he was 200 yards short of completing the second 1km run, he did not break his own record. His final words, from his cockpit radio, were: 'Full power... Tramping like hell here... I can't see much... and the water's very bad indeed... I'm getting a lot of bloody row in here... I can't see anything... I've got the bows up... I'm going....oh!'

The current world water speed record, of 317.6mph, is held by the Australian Ken Warby.

THE SPEED OF LIGHT

ARGUMENTS over the speed of light had raged for centuries until English vicar and amateur astronomer **James Bradley** discovered the 'Aberration of Light' in 1728. The Aberration of Light is important, but quite dull; the point is that it enabled Bradley to prove beyond all doubt – and for the first time in history – that the Earth was moving through space and that the 'Heavens' did not revolve around us. It also allowed him to calculate the speed of light, which he did to within 1% of the currently accepted figure of 299,792,458 metres per second (or 670,616,629 mph). Prior to this, most scientists had considered it to be infinite. Bradley, who was born in Sherbourne, Gloucestershire, in 1692, became Astronomer Royal and died in 1762.

SQUASH

FIRST played at Harrow public school in 1817, squash evolved from the game of 'rackets' played at Fleet Debtors' Prison and is now played in 175 countries by an estimated 15 million people.

STAINLESS STEEL

HENRY Brearley from Sheffield invented stainless steel pretty much by accident in 1912. He was actually trying to make a better gunbarrel steel – one which would be less prone to wear out under the stress of firing – and experimented by adding chromium and varying the carbon content. He called the resulting product 'rustless steel' and sought a US patent in 1915, when the *New York Times* announced the arrival of the new 'non-rusting, unstainable and untarnishable' metal.

STAMPS

THE Royal Mail dates back to 1516, when Henry VIII established a 'Master of the Posts' to distribute Royal and Government documents. The service was first made available to the public by Charles I in 1635, although it was the person *receiving* the letter who had to pay the postage.

The 'penny post' was introduced in England in 1680 by **William Dockwra**, covering only letters being mailed within and to addresses in central London. For two centuries, letters were marked with ink to record the date that they had been mailed. Then, in 1840, **the world's first adhesive postage stamp** was launched. The Penny Black bore Queen Victoria's head on a black background and allowed a letter weighing no more than half an ounce to be sent anywhere in the country for a penny.

Although the stamp was certainly a British invention, there is debate as to whether it was down to a Scot or an Englishman. Most people credit **Rowland Hill**, (1795-1879), a school teacher and postal reformer from Kidderminster who had proposed the use of postage stamps in a document published in 1837; Hill was knighted for the invention in 1860. But Dundee bookseller **James Chalmers** (1782-1853), from Arbroath, Scotland, always argued that he had come up with the idea three years before Hill. He was largely ignored, though he was awarded a silver jug and 50 sovereigns at a public meeting in Dundee in 1846. After his death, Chalmers' son Patrick had his father's headstone engraved with the claim that he was the 'Originator of the adhesive postal stamp' which 'has been adopted throughout the postal systems of the world'.

British stamps are the only ones which do not have their country of origin printed on them, in recognition that we were the first to issue stamps.

STEM CELLS

BRITISH scientists grew **the world's first transplant organ using a patient's own stem cells**.

Claudia Castillo, a 30-year-old Columbian whose airways had collapsed following tuberculosis, needed her windpipe replacing to save her left lung. **Martin Birchall**, a professor at Bristol University, led a team which created a new windpipe for her in 2008. Cells lining her windpipe and stem cells from Miss Castillo's bone marrow were harvested and used to develop new cartilage cells at Bristol. Then a donor windpipe from a patient who had recently died was stripped of its cells with strong chemicals – this 'skeleton' was coated with the new cartilage. After four days in the lab, it was ready for transplant. The operation took place in Spain (Italian scientists were also involved), and Miss Castillo is now in perfect health.

Professor Birchall said: 'I reckon in 20 years' time it will be the commonest operation surgeons will be doing. We are on the verge of a new age in surgical care.'

Stem cells are master cells which can divide to provide copies of themselves and other types of cell. They are so versatile that they could potentially be used to repair and replace any damaged human tissue. By using the patient's own stem cells, the need for powerful anti-rejection drugs is removed.

British scientists lead the field globally. **The world's first 'new' heart valve** was grown at Imperial College London and Harefield Hospital. It is hoped that lab-grown valves will be commonplace within three years. A cure for deafness is being developed at the University of Sheffield, researchers at King's College London and Nottingham University believe they are close to repairing brain damage caused by strokes, and new treatments for Parkinson's Disease, infertility in men and strokes are being developed at Newcastle University.

In many cases, stem cells are taken from aborted embryos, making the science highly controversial for some people. But a team headed by **Ian Wilmut** (see *Cloning*) at Edinburgh University has found a way of creating stem cells out of adult skin cells – this will remove the need for embryos to be used.

STEREO SOUND

A BRILLIANT British electronics engineer vastly improved the telephone, invented stereo sound and pioneered high definition television – all in a tragically short career which ended in a wartime air crash while testing a new form of radar. **Alan Blumlein** (1903-1942) was only 38 when he died in 1941, but he left an enormous legacy of 128 patents, many of which still shape today's world.

While working for an early telecoms company, he developed a way of making long-distance phone calls much clearer; his technology lasted until the advent of digital communications decades later, and he was awarded a year's salary as a bonus. His development of stereo – he called it 'binaural sound' – came after a visit to the cinema with his wife, Doreen. At the time, films had 'mono' sound tracks and only one set of speakers, and Blumlein was irritated by watching an actor talk on one side of the screen while the voice appeared to originate on the other. He was so far ahead of his time that it was 1958 before his peers recognised his achievement. He also made many advances in television, including the first forays into high-definition.

When WWII broke out, he joined the RAF and quickly contributed to the design of a new Air Interception radar which allowed our planes to pick off German aircraft in the night skies over Britain (ironically, given that Blumlein's father was a German who became a naturalised Briton). He made an invaluable contribution to a navigation radar system called 'H_2S' and related equipment used for locating enemy ships and U-boats – in a 1943 radio broadcast, Hitler complained that growing Allied successes against his Atlantic submarine fleet were due to 'a single technical invention of our enemy'. Sadly, Blumlein had died the previous year, when a Halifax bomber being used to test H_2S crashed in the Wye valley, killing all aboard. He left his wife and two sons, Simon and David.

STRANGE CUSTOMS

WE'VE given the world cheese rolling, shin kicking and bog snorkelling, among many other slightly odd customs, pastimes and traditions.

Cheese rolling dates from at least 200 years ago, and involves hundreds of whooping lunatics chasing a large round cheese down Coopers Hill near Cheltenham. Those who have studied the video footage claim that the cheese can reach 70mph, and serious injuries are relatively common. Whoever gets to the bottom of the hill first wins the cheese, which they can eat in hospital.

Shin kicking is also a quaint Gloucestershire sport, dating from 1612. Held as part of the annual the Cotswold Olympicks at Dover's Hill near Chipping Camden, the rules are quite simple: you grasp the shoulders of your opponent and attempt to kick his (or her) shins. According to organiser Robert Wilson, players of yesteryear hardened their shins using hammers to prepare for the event. Crowds of 30,000 would turn out to watch in the 1830s (though that *was* before telly).

In Wales, they invented bog snorkelling as late as 1985. Competitors swim two lengths of a filthy, 60-yard trench cut through the Waen Rhydd peat bog, near Llanwrtyd Wells. Competitors must wear snorkels and flippers, and rely on flipper power alone.

There's far, far more – toe-wrestling, coal-carrying, snail-racing, the World Nettle Eating Championships, six-a-side football matches held in (and I do mean 'in') the River Windrush at Bourton-on-the-Water, Cambridgeshire's World Pea Shooting Championships, the Birdman Festival (where human-powered flying machines are launched from Worthing pier with predictably disastrous and hilarious results) – but I really don't have the strength to carry on.

STREET LIGHTING

A STUBBORN and short-tempered Scot unveiled **the world's first streetlights** in Britain in 1794.

William Murdock was working in Cornwall for the steam engine pioneers James Watt and Matthew Boulton (see *Industrial Revolution*), when he collected the gases given off when coal was heated and used them to light his house in Redruth – much to the concern of his neighbours, one of whom was Richard Trevithick (*ibid*).

In 1802, he put on an outside lighting display at Boulton's Birmingham foundry – this must have amazed onlookers, as until then the only form of artificial lighting the world had seen was from flickering candles or smoky oil lamps. By 1807, London's Pall Mall became **the first street in the world to be lit at night**, and over the next four decades gas lighting spread to every town and city in Britain.

Over 6ft tall and broad-shouldered, Murdock also made the first portable gas lantern, a pig's bladder filled with coal gas which was squirted through an old tobacco pipe and lit – imagine bumping into him at night in the early 1800s.

He was born near Cumnock, East Ayrshire, in 1754. The third of seven children, and the first son in his family to survive beyond infancy, he was an excellent mathematician as a schoolboy. Determined to get a job with James Watt, as a young man he walked all the way from his home in Scotland to Birmingham to present himself. This clearly did the trick and he was taken on. As well as developing Watt and Boulton's steam engines, he created the first gas meter, a door bell which worked on compressed air and a system of tubes which used compressed air to send messages between rooms in large buildings. This was widely adopted in offices: Harrods used one until the 1940s. He also revolutionised the brewing industry with a replacement for isinglass, an expensive substance found in the bladders of Russian sturgeon which was used to clarify beer. Murdock's replacement, made from cod, was far cheaper and the Committee of London Brewers paid him £2,000 for the rights to his invention.

It's worth making mention of the heartbreak which marked his life, because it shows how hard mere existence was back then. While in Cornwall Murdock fell ill with malaria and was nursed back to health by a local girl called Ann Paynter. The nursing must have involved some unusual techniques, because Ann fell pregnant. Under pressure from Matthew Boulton – who felt responsible for his employee's behaviour – and the girl's father, they were married. In August 1786 she gave birth to twins but they were dead within two years, to the couple's great distress. Ann produced two more sons, but herself died shortly after the arrival of the second. Nowadays, tragedy on that scale would make the national newspapers; in the 18th century it was commonplace.

Murdock lived on for many years after his wife's demise, his determination, hard work and ingenuity being richly rewarded. When he died in Birmingham in 1839, aged 85, he was a rich man; he left an estate worth more than £30,000, £24 million today.

SUBMARINES

DESPITE the fact that Britain's rivers and seas are generally pretty cold and uninviting, we led the way in the early technical development of underwater travel.

The world's first practical submarine was invented by a Gravesend pub landlord called **William Bourne** in 1578.

Bourne, a former Royal Navy gunner, designed a wooden vessel covered in waterproofed leather which could descend by decreasing its volume and re-inflate to return to the surface again. Wooden screws would be used to expand and contract the leather shell, and oars would be used to propel it underwater. The principle was sound but unfortunately he never got around to building it, perhaps realising that he would be better off spending his time behind the bar in his pub.

It wasn't until 150 years later that **Nathaniel Symons**, an English carpenter, built and demonstrated a working example following Bourne's plan. In 1729, he spent 45 minutes underwater as a crowd watched, before rising back to the surface and passing around a hat. The onlookers don't seem to have been all that thrilled:

the submarine historian Captain Brayton Harris of the US Navy says, 'One man gave him a coin.'

In 1629, meanwhile, London had also seen **the world's first recorded underwater journey** – by a dozen men travelling down the Thames submerged to a depth of about 15 feet. (Admittedly, this craft had been built by a Dutchman, Cornelius Drebbel, but he was the court inventor to King James I, and his machine was built here.)

We also built **the first serious steam-powered sub**, *The Resurgam*, in 1879. In keeping with the unlikely backgrounds of British submariners, this was the brainchild of a vicar, the **Rev George Garrett**. *The Resurgam* was built at Birkenhead for £1,500 (£777,397.84 in today's terms) but she sank while being towed to Portsmouth.

Unfortunately, having got ahead of the crowd back in the 16th century, the sad truth is that we never really capitalised on our innovation as we should have, mostly because suspicions persisted in the higher echelons of the Royal Navy that there was something slightly ungentlemanly about underwater warfare, with its whiff of low-down sneakiness and absence of fair play. In 1900, Rear Admiral A K Wilson insisted that the submarine was 'underhand, unfair, and damned un-English'. Their crews should be treated 'as pirates' and hanged if captured, he said. The Admiralty, too, was sniffy. 'They are the weapons of the weaker power,' read a report. 'They are very poor fighting machines and can be of no possible use to the Mistress of the Seas.'

Perhaps this is unsurprising, given that we had only recently stopped sending our soldiers into battle dressed in bright red tunics to give the enemy a better target.

Attitudes changed gradually, and the Royal Navy Submarine Service acquitted itself very well in both world wars. But other nations had moved ahead, particularly Germany. Churchill admitted after the end of WWII: 'The only thing that really frightened me was the U-Boat peril.'

Incidentally, the Royal Navy submarine Captain Roger Bacon invented the sub's periscope.

SUBMARINES – DEPTH CHARGE

HAVING invented the first submarine, it was perhaps inevitable that we would also come up with the first depth charge.

An underwater bomb that exploded at a pre-set depth, it was invented by Royal Navy boffin **Herbert Taylor** in Portsmouth in 1916.

Very shortly afterwards Germany became the second nation to discover the depth charge, when their U-boat *U-68* was sunk by the new weapon. It was an important development; German subs patrolling the Atlantic posed a major threat to merchant shipping and military resupplies coming across from the United States. (In 1915, the sinking of the civilian liner *Lusitania* off Ireland – she was torpedoed, with the loss of 1,198 lives, after being mistaken for a troopship – showed the scale of the threat.)

At first, they were relatively unsophisticated. Their barrel-like shape meant they sank slowly, and they had only two depth settings, 40ft and 80ft, and two explosive charges, either 120lb and 300lb of TNT or amatol (the smaller charge was intended to avoid destroying the ships themselves where they were too slow to escape the blast radius of a 300lb bomb). Surface vessels would try to bracket the sub by dropping charges to explode at different depths. It was a hit-and-miss affair, but the Second World War saw more powerful weapons, streamlined so they sank quicker and to greater depths. Sonar had also arrived, making the business of locating and attacking the enemy more scientific (though German submarines had also evolved, to the point where they could survive multiple depth charges: the *U-427* is said to have survived 678 depth charge blasts in April 1945).

As the threat posed by submarines escalated in the Cold War years – a single Soviet ballistic missile sub had the potential to destroy all of the UK's major cities – we even developed nuclear depth charges whose enormous explosive power guaranteed a successful hit on the enemy. Anti-submarine torpedoes have now replaced the depth charge.

SUBMARINES - TORPEDO

IT comes as little surprise, further, that the torpedo was also invented by a Briton, **Robert Whitehead**. Born in landlocked Bolton in 1823, Whitehead developed the first self-propelled torpedo in 1866 while working in Trieste, northern Italy, for a company employed by the Austro-Hungarian navy. By 1870 his compressed air-powered weapon could hit a target 700 yards away at a speed of 10mph. The Royal Navy's Submarine Museum website quotes Admiral H J May as saying, 'But for Whitehead, the submarine would remain an interesting toy, and little more.' Ironically, Whitehead's granddaughter married one Georg Ludwig von Trapp in 1911 – a man who went on to become a Germany Navy U-boat ace in WWI.

SYNTHETIC FIBRES

AS early as 1664, the brilliant polymath Robert Hooke began experiments to find a man-made fibre that would be better than silk, though he never succeeded. Lightbulb inventor Joseph Swann did – the carbon fibres he used as filaments were spun into artificial silk and shown at the Exhibition of Inventions in 1885 – but he was too busy experimenting with electricity to follow it up.

So fabrics continued to be made from natural materials until British chemists **Charles Cross** and **Edward Bevan** came up with a cheap alternative – a thread made from wood pulp – in 1892. Cross, from Brentford, Middlesex, and Bevan, who was born in Birkenhead, invented viscose rayon (cellulose acetate).

In 1894, they patented their manufacturing method, making it **the world's first practical man-made fibre**. (The French scientist Hilaire de Chardonnet had invented a viscose fibre, cellulose *nitrate*, slightly earlier using methods virtually identical to Swann's, and demonstrated his 'artificial silk' at the Paris Exhibition in 1889, but the chemical composition of the British fabric made it less flammable and easier to spin into thread). By the 1920s, viscose rayon had replaced cotton and wool as the material most likely to be used in women's underwear and stockings.

The next British triumph – well, it seemed like it at the time – was the invention of polyester. This was a cheap, strong, durable, non-crease material made from the same polymers found in plastic bottles. Chemist **John Whinfield** (1901-1966), from Sutton, Surrey, developed it in a research lab in Manchester, helped by Scot **James Dickson**. It was patented in 1941, but kept secret during the war. It was later marketed as 'Terylene' and formed the basis of drip-dry and easy care clothing and polyester was (and in some cases still is) used in everything from sportswear to drinks bottles.

THE TANK

LEONARDO da Vinci imagined them, and the sci-fi writer H.G. Wells wrote about them, but **the world's first tank** was British.

A **Major General Sir Ernest Dunlop Swinton** was the driving force behind its development, after conceiving the idea in October 1914 while driving in France. The first tanks appeared on the Somme in 1916, having been developed by the Royal Navy at Sir Winston Churchill's request as a means of overcoming the stalemate of the trench warfare of WWI.

The tanks of that war – the Mark I to Mark IV – were slow (with a top speed of around 4mph), cumbersome, lightly armoured (around half an inch) and lightly armed, with 57mm 'six-pounder' guns and machine guns. They were rhombus-shaped, with no turret, and were generally quite ineffective – mostly because the Army hadn't yet worked out how to use them. As the Germans discovered in WWII, tanks are best employed as a means of punching quickly and aggressively through enemy lines with infantry in close support. Early British tactics often saw unsupported tanks getting bogged down in mud.

The Germans and the Soviet Union developed the best tanks of WWII – the Tiger and the T34 respectively – but **the world's best modern-day tank** is thought by many to be Britain's Challenger II, a 62-tonne monster with a top speed of almost 30mph, reactive (and top secret) 'Chobham' armour which will stop almost any projectile and a 120mm gun which can hit targets five miles away with astonishing accuracy. During the 2003 invasion of Iraq, Challengers regularly took on and destroyed multiple Iraqi Soviet-built tanks at no loss to themselves.

TARMAC

ALL roads may lead to Rome, but they were paved by two Britons.

The first was **John Loudon McAdam**, who was born in Ayr in Scotland in 1756 into a wealthy family which lost almost everything in a banking collapse. John was sent to America to live with his uncle, a merchant in New York. He prospered, and returned to Scotland in 1783, buying a small estate in Ayrshire. He built several roads on his land and became interested in their construction, but it wasn't until 1801 – after he and his family had moved to the west of England – that his career as a road builder began. He had taken a position as surveyor to the Bristol Turnpike Trust, which was in charge of maintaining the roads in that area as well as collecting tolls from their users. The quality of the roads varied from passable to boneshaking and McAdam decided to try and improve them.

He took a given track and drained and levelled it. Large stones would form a base, and then small, irregular-shaped rock chippings – plum-sized, and weighing no more than 6oz – would be laid on top to a depth of 10 inches, before being compressed. As traffic rolled over the top, it tended to consolidate the surface more, creating a strong surface. McAdam also left a slight camber of an inch in every yard from the centre of his roads to help water run off.

This was the first major advance in road construction since Roman times and was widely adopted from the 1820s onwards. It became known as the *Macadam Method*, making him among the few whose insight leads to their name becoming part of the language. He died in 1836.

The second Briton in the story is a Welshman, **Edgar Hooley**. Born in Swansea in 1860, by 1889 he was chief surveyor to Nottinghamshire county council. Over the next few years, he realised that McAdam's methods were not sufficient for the increased amount of traffic on the roads. As well as creating dust in dry weather, they were prone to erosion after heavy rain, particularly when combined with the sucking action from the pneumatic rubber tyres on new motor vehicles. He discovered a way of combining tar, iron slag and granite chippings to produce a hard-wearing surface. He called this 'tarmacadam' and Nottingham's Radcliffe Road, south of the city, was **the first tarmac road in the world**. He patented the system and set up a company – now itself called Tarmac, and still in business today, when the McAdam-Hooley method is used everywhere. There is no definitive answer as to how many miles of roads are thus covered, but the *CIA World Factbook* estimates that in 2008 there were 68,937,575 km (42,835,823 miles) of paved roads in the world.

On the subject of roads, special mention should also go to **Thomas Telford**. Born the son of a poor Scottish shepherd in 1757, he became one of the greatest engineers the world has ever seen and the first president of the Institution of Civil Engineers. The so-called 'Colossus of Roads' developed similar methods to McAdam and laid highways all around the UK, as well as building hundreds of miles of canals and dozens of bridges. Among his most famous works were London's St Katharine's Dock, the Menai Suspension Bridge linking mainland Wales to Anglesey, and the stunning Pontcysyllte Aqueduct which carries the Llangollen Canal over the River Dee near Wrexham. Completed in 1805, it is the longest (1,0007ft) and highest (126ft) aqueduct in Britain.

THE TAXI

THE first taxis were 'Hackney carriages' pulled by a pair of horses; introduced in the 1650s, they toured the streets of London looking for passengers to pick up. In 1834, **George Hansom** improved the design of the Hackney carriage, producing a lighter, low slung version which could be pulled by one horse. The Hansom Cab was born ('cab' being short for cabriolet) and they had soon spread throughout the land and as far afield as New York and Sydney. In the early 20th century, they were gradually replaced by the distinctive, motorised 'Black Cab'. There have been several versions, but the 'classic' London taxi was the Austin FX4, made by Carbodies in Coventry between 1958 and 1997. This was replaced late in the last century by the similar-looking TX range produced by London Taxis International.

TEA

A NICE pot of tea is the British answer to any disaster – we brew 130 million cups a day. We didn't discover it, but we did **give tea to the world**. We did this by an act of daring daylight robbery.

For centuries, tea came primarily from China, where the mysteries of its cultivation and ways of drying and using the leaf were controlled as strictly as any modern military secret. If you wanted tea, you bought it from China, and in the 1700s we imported it by the ton, in huge ships called tea clippers which took more than a year to reach England.

In 1788, **Joseph Banks** (see *Kew Gardens*), an Eton-educated explorer and botanist who had sailed aboard *Endeavour* with Captain Cook on the first of his epic voyages to the Pacific, suggested tea growing might be possible in British-controlled north east India. The only problem was, how to obtain the plants and know-how? The Chinese punished smuggling with summary decapitation.

In the 1840s, **Robert Fortune**, a botanist and adventurer born in Kelloe, Berwickshire, undertook several highly hazardous journeys into the closed country, disguised as a Chinaman, and escaped with thousands of samples and an intimate knowledge of tea horticulture. He took 20,000 tea plants to Darjeeling after they had been propagated in London. They grew beautifully and, even better, the leaf grown in India was stronger-tasting and much more to our liking. By 1900, only 5% of tea imports to Britain were from China. Robert Fortune had destroyed the Chinese tea monopoly forever.

Funnily enough, coffee was here earlier; tea didn't arrive until the 1660s when King Charles II married the Portuguese Catherine of Braganza. Tea chests were part of her wedding dowry and she introduced it to the Royal household and the aristocracy. Fifty years later, Queen Anne started drinking tea with her breakfast, rather than the customary beer. Afternoon tea was invented by the Duchess of Bedford in the early 1800s. She complained that she felt faint with hunger if she didn't have something to nibble between lunch at 1pm and dinner at 7pm.

TEASMADE

DERBYSHIRE inventor **Samuel Rowbottom** was granted the first patent for an automatic tea-making machine in 1892. He rigged up an alarm clock to a gas ring and a pilot light. The gas ignited at the set time and boiled a pot of water. Steam pressure forced the water out through a tube and into a teapot.

A prototype was made but it was never produced commercially. It was not until 1936 that the electric automatic tea maker – known as the Teasmade – was mass-produced by Surrey-based Goblin. As well as a kettle, it featured a lamp with a shade, an alarm clock and, later, a radio. At one point, two million households owned a Teasmade.

THE TELEGRAPH

BEFORE the arrival of the telegraph, the only ways of communicating over distance were the ancient burning beacon, lit on high ground, or semaphore (flags waved in varying orders to send a given message). The British patented **the world's first commercially-viable telegraph**.

Built by **Sir William Fothergill Cooke** (1806-1879) for the Great Western Railway, it was unveiled on April 9, 1839, and ran for 13 miles from Paddington station to Hillingdon in west London. However, it might have arrived much earlier, but for the mulish attitudes of some within the hierarchy of the Royal Navy.

Francis Ronalds, born in 1788 into a family of cheesemongers, had set up an eight-mile telegraph in his own garden in Hammersmith in 1816. The principle was very simple; the sender at one end of the wire could interrupt the flow of electrical current by pressing a 'key' and this break could be detected by the receiver. A pre-arranged code could be used to send a simple message. The international Institution of Engineering and Technology's website explains how Ronalds wrote in 1816 to offer his invention to John Barrow, Secretary to the Admiralty. Barrow replied that 'Telegraphs of any kind are now wholly unnecessary.' In a pamphlet published in 1823, Ronalds listed some of its potential uses. 'Why should not government govern at Portsmouth almost as promptly as in Downing Street?' he asked. 'Why should our defaulters escape by default of our foggy climate? Let us have Electrical Conversazione offices communicating with each other all over the kingdom if we can.'

He was ignored, but his ideas were later taken up with enthusiasm by slightly brighter people than Mr Barrow and Ronalds was knighted in 1870, three years before he died, for his 'early and remarkable labours in telegraphic investigations'. The first demonstration of electrical telegraphy was in 1810 by the German inventor Samuel Thomas von Sömmering.

THE TELEPHONE

'MR Watson, come here. I want to see you.' These are not, in themselves, particularly dramatic words – but then it was the medium, rather than the message, which makes them among the most famous ever spoken. They were said by a Briton, **Alexander Graham Bell**, when he made **the world's first telephone call**.

Bell was born in South Charlotte Street, Edinburgh, in 1847, a time when the only way to talk to someone was face-to-face. By the time he died in 1922, there were an estimated 13 million telephones in use worldwide.

He came from a family interested in sound and communication. His grandfather and father, both named Alexander, were experts in elocution and developed cures for speech impediments, and his mother, Eliza, was deaf. In his 20s, Bell crossed the Atlantic and worked as a teacher, both of the deaf and of those who taught the deaf. Part of his work involved investigating speech vibrations and sound waves and their effects on the membranes of the ear. In June 1875, while drawing on his expertise in these areas to work on a 'musical telegraph' which could transmit and receive multiple messages over the same wire at once (standard machines could only deal with one 'dot-dot-dash-dash' message at a time), he and his assistant, Thomas Watson, found that they were transmitting between two machines the actual sound of a metal reed being vibrated. Realising that this offered far more possibilities than *any* telegraph, Bell worked feverishly on constructing and testing a variety of improved transmitters and receivers until, on March 10, 1876, he succeeded in sending his famous phrase to Watson, 20ft away in another room.

He was 29 when he filed his patent, which covered 'the method of, and apparatus for, transmitting vocal or other sounds telegraphically'. The telephone was first demonstrated at an exhibition in Philadelphia three months later. The emperor of Brazil, Dom Pedro de Alcântara, put the receiver to his ear and proclaimed: 'My God, it talks!' Lord Kelvin (of *Transatlantic Cable* fame, and much more) was also at the exhibition; he described it as, 'The greatest marvel hitherto achieved by the electric telegraph.'

The Bell Telephone Company was set up in 1877, and by 1886 150,000 people in America were connected. Bell, who refused to have a telephone in his own study, also made **the world's first long distance call** in 1915 when he spoke from New York to Watson, 3,000 miles away in San Francisco.

The telephone wasn't his only invention. His first was a wheat de-husking machine, made when he was 12, and he was most proud of his 'photophone' – a method of transmitting sound using a light beam, an early form of fibre optic technology. Obsessed with flight, he also held several patents for 'aerial vehicles', and in 1919 invented a hydrofoil capable of 70mph. There was also a gramophone, a metal jacket to assist breathing – later developed into the iron lung used for polio victims in the 1950s – an audiometer to detect hearing problems, an iceberg detection device and an early form of air conditioning which used fans to blow air across huge blocks of ice. After American President James Garfield was shot, Bell came up with a metal detector to try and find the bullet. (Unfortunately, Garfield was lying in a metal-framed bed which made the machine malfunction.) He wrote about the possibility of harnessing the Sun's energy in solar panels and predicted that in the future telephones would be wireless, imagining a time when men would have 'coils of wire about their heads coming together for communication of thought by induction'.

When he died of diabetes in Canada, aged 75 in August 1922, all telephones in America were silent for one minute in honour of their inventor.

THE TELESCOPE

THE first practical telescopes were made by Britons, and their impact was enormous: they helped scientists to disprove religious claims that the Earth occupied a special and central place in the Universe.

The earliest telescopes, consisting of two lenses at either end of a tube, had been made in Holland in the early 1600s. But while they worked well enough to allow Harriot and Galileo to gain the first close-up views of the Moon, and to see some of the solar system's planets, they were still rudimentary devices. Lens improvements helped, but the physics involved in 'refracting' telescopes of this kind meant that high-quality images were impossible – 'chromatic aberration', a phenomenon by which the component colours in an object are focused at different points, led to blurred and indistinct views. This could be ameliorated by placing the lenses a significant distance apart, and telescopes as long as 140ft were proposed. But this created other problems – it then became next to impossible to keep the lenses aligned, because the technology to make the tubes rigid enough did not exist.

The first real breakthrough in the technology was made by the brilliant British astronomer and mathematician, **James Gregory**. Born in Aberdeenshire in 1638, Gregory was 25 years old when he designed the reflecting telescope, which employed mirrors within the tube to cure the issues refractors posed by collecting and focusing light. Unfortunately, while he was a genius thinker – rated second only to Newton by some – he was no engineer, and had no idea how to build it. Indeed it wasn't built until a decade later, by Robert Hooke using Gregory's notes. (Gregory, a great mathematical pioneer, solved major problems in algebra, calculus and geometry but suffered a stroke in 1675 and died a few days later, aged just 36. Who knows what he might have accomplished given more time?)

In 1668, between the Gregorian telescope's design and its construction, **Sir Isaac Newton** actually built the **first successful reflecting telescope**. Newton, who ground and polished the mixed tin and copper mirrors himself and devised a means of measuring their optical quality, used slightly different methods but the same basic principle. (There is a lovely sketch, by Newton himself, of his device at www.antiquetelescopes.org.) Most optical telescopes today use reflecting technology of the sort pioneered by Gregory and Newton.

The importance of telescopes was huge because they proved that the Polish astronomer Nicholas Copernicus was right when he suggested that our planet revolved around the Sun – rather than the reverse – in 1543. Decades later, 'heliocentrics' were still being burned at the stake as heretics, and Galileo himself was placed under house arrest for life in 1633 for the same 'crime'. But with the astronomical cat out of the bag, power began to shift slowly away from theologians to rationalists.

TELESCOPIC SIGHT (AND THE MICROMETER)

A SPIDER was an astronomer's unlikely accomplice in his invention of **the world's first telescopic sight** in the 1630s.

Leeds-born mathematician and scientific instrument maker **William Gascoigne** was staring at the night sky when a web landed on the guide of his telescope. He noticed that the thread stood out sharply. When another strand formed a cross at the centre of the lens they provided an ideal measuring point and allowed him to aim the telescope more precisely.

Gascoigne's most famous invention is the micrometer – a precise measuring device used by engineers. He intended it for use in astronomy, to measure the size of, or distance between, the planets and stars. It consisted of pointers which could be seen through the telescopic lens. A screw was turned to close or open the pointers, and he could enclose the diameter of the Moon, or the distance from a given star to its neighbour, within them. A measuring dial attached to the screw showed how far apart the pointers were for each fraction of a turn. The device enabled Gascoigne to work out with greater accuracy the distances and sizes in astronomy. Its successors are still in use today. Sadly, he was killed in the Civil War battle of Marston Moor, in Yorkshire, on July 2, 1644, aged just 32. Again, we can only speculate as to what he might have achieved had he lived longer.

THE TELEVISION

IT'S a moot point whether the gogglebox is a revolutionary force for good, or a dreadful contraption which is gradually turning the human race into morons. Either way, the British **gave the television to the world**.

The credit for inventing the first TV goes to a soap salesman from Dunbartonshire called **John Logie Baird**, though it was a mechanical system quite unlike the far superior electronic televisions which very soon superseded it.

Born a clergyman's son in Helensburgh in 1888, Baird was always interested in technology – as a young lad, he rigged up a telephone exchange to connect his bedroom with those of his friends across the street, and he also set up a petrol-driven generator to power electric lights for his parents. He attended Glasgow's Royal Technical College, where he took a Diploma in Electrical Engineering, and then enrolled at the city's university on a BSc course, although he left before graduating. In 1916, he joined the Clyde Valley Electrical Power Co, but he left after the end of WWI, intending to make his own way selling household products like boot polish, fertiliser and the aforementioned soap. At the back of his mind, though, was a far grander scheme – the idea of creating a 'radio' that broadcast pictures.

Baird's health was never very good, and in 1922 he moved to the south coast of England where he hoped the air would be better. It was there that he applied himself to creating a television. In the best traditions of Newton, he stood on the shoulders of many others to achieve his goal. One of them was 'fax' inventor Alexander Bain; another was Willoughby Smith, a Norfolk electrical engineer who had discovered the element selenium's sensitivity to light, a finding which led to the invention of photoelectric cells used in television; a third was the German electrical engineer Paul Gottlieb Nipkow, who had developed a method of scanning pictures which allowed them to be broadcast. (There were others, too.) With little funding, Baird built the first television from odds and ends – including an old hatbox, some bicycle light lenses, a tea chest and a Nipkow scanning disk – and succeeded in transmitting images on to a screen a few feet away. (There is a working model of his apparatus in the Science Museum.)

In February 1924, now working in a London lab, he transmitted moving silhouette images as a correspondent from the *Radio Times* watched. The first public demonstration came at Selfridges in March 1925. At this stage, the picture refreshed only five times per second, so the images were jerky and indistinct. By the following January, Baird had improved the 'scan rate' to 12.5 pictures per second and showed his invention off to members of the Royal Institution and a *Times* reporter. This is regarded as **the world's first demonstration of a true TV system** which was able to broadcast live moving images of reasonable quality. In the same year, he transmitted **the world's first long-distance television signal** over 438 miles of telephone line between London and Glasgow.

In 1928, he produced **the world's first colour transmission** and his Baird Television Development Company made the first transatlantic television transmission, from London to New York, and the first experimental television broadcast for the BBC. The 1931 Epsom Derby was shown at a cinema in central London, to the amazement of viewers. In the early 1930s, ignoring the fact that no-one owned a telly, the BBC began public service TV using Baird's technology. But it was short-lived. From the mid-1930s, his mechanical TV was gradually replaced by electronic systems which used the superior cathode ray systems developed by the Russian inventors Vladimir Zworykin and Isaac Shoenberg at London's EMI-Marconi. Baird's system was finally dropped by the BBC in February 1937; by then, there were broadcasts in London for an average of four hours a day and there were roughly 400 television sets in the UK.

Although he is best-known for television, Baird was actually a scientific polymath who recorded a staggering 178 patents over a 20 year period. Some are forgotten, but others were crucial in the development of radar, 3D TV, fibre optics and infra red night sights.

He died after a stroke in June 1946, and never lived to see today's era of mass TV. So what would he have made of *Big Brother* or *I'm A Celebrity, Get Me Out Of Here?* His son Malcolm, a scientist now living in

Canada, said in 2006 that if his father had known how things would turn out, he would never have invented it.

TELEVISTA

ARCHIBALD Montgomery Low (1888-1956) was a prolific inventor with a talent for seeing into the future. Born in Purley, near London, the engineer's son predicted the use of mobile phones at a time when only a few wealthy families had *landline* telephones, developed **the world's first pilotless plane** and, in 1914, discovered how to transmit moving images over a wire. He called this process 'TeleVista' and he invented it 11 years before Baird gave his first public demonstration of television at Selfridges. However, the images it produced were inferior to those Baird would create, and Low's attention was taken elsewhere by the outbreak of WWI.

He turned his attentions to remote-controlled weapons, working on ways in which pilotless aircraft filled with explosives might be turned into flying bombs. The rather staid British Government and military failed to appreciate the possibilities but the Germans were more alert: in 1915, they made two unsuccessful assassination attempts on his life. An agent tried to shoot him through his laboratory window and when that failed a friendly chap with a strange accent offered him a cigarette. Low became suspicious and did not smoke it, which was just as well since it was laced with strychnine. The Germans improved upon his guidance system and used it in their V1 flying bomb (also known as the 'doodlebug'), which rained death onto England and Western Europe during WWII. Low also invented pre-selecting gears and, with less effect on posterity, the whistling egg-boiler.

TENNIS

ARGUMENTS as to the history of tennis are murky and usually revolve around the old favourite – who was the first to play a 'tennis-like game'? What's beyond doubt is that the modern game was developed here. The first tennis club was set up in Leamington Spa in 1872 by **Major Harry Gem** and a Spanish friend, Augurio Perera, and the first Wimbledon championships were played in 1877. In those days, British players were almost certain to win, of course. Indeed, the youngest Wimbledon champion was 15-year-old Cheshire girl Lottie Dod, who won the first of five titles in 1887. She was also the first woman on court to display her ankles, wearing a slightly-shorter-than-ground-length dress to give more freedom of movement.

We also developed table tennis – otherwise known as ping pong or (by Boris Johnson) 'whiff-whaff', this was originally an evening diversion for rich and obviously bored Victorians, who would use random household items to hit a rounded cork over a 'net' of books placed on a dining table. The first celluloid balls were employed by Englishman James Gibb in 1901, swiftly followed by rubberised bats. Not long after this, the England footballer Max Woosnam apparently defeated Charlie Chaplin while playing with a knife; this would have been like Wayne Rooney beating Brad Pitt at snooker with a rake, something I would pay to watch.

TEST TUBE BABIES

THE pioneering work of two British fertility experts led to **the world's first test tube baby**.

John and Lesley Brown, a couple from Bristol who had been trying unsuccessfully for a family for nine years, were delighted when their daughter Louise arrived in 1978. But their dream would not have been possible without the brilliance of consultant gynaecologist **Patrick Steptoe** and **Professor Robert Edwards**.

Steptoe and Edwards removed a single egg from Mrs Brown's ovary, fertilised it with her husband's sperm in a glass dish ('*in vitro fertilisation*') and implanted the resultant embryo back into her womb. Louise Joy Brown was delivered by caesarean section at 11.47pm on July 25 at Oldham General Hospital, Lancashire. Her mother said: 'It's a miracle. Louise is truly a gift from God.'

Around four million babies worldwide have since been born thanks to this technique. In 1980, the two men founded Bourn Hall Clinic in Cambridgeshire, now a world centre for the treatment of infertility. Mr Steptoe died in 1988, aged 74, and Prof Edwards, now in his 80s, recently told how he relayed news of the clinic's 1,000[th] IVF birth to his friend and colleague as he lay dying. 'I'll never forget the look of joy in his eyes,' he said.

As early as 1891, Cambridge University professor **Walter Heape** (1855-1929) had successfully taken embryos from one rabbit and successfully placed them in the uterus of another – **the world's first embryo transfer** and a forerunner of later reproductive science.

TEXT MESSAGES

A BRITISH engineer sent **the wrld's 1st txt msg** 2 a mobile phone on December 3, 1992. It was the start of a global phenomenon – a billion texts are now sent each week in the UK *alone*. That first message was not very profound, but then few of them are. It said simply, 'Merry Christmas', and was sent by 22-year-old **Neil Papworth**, a communications engineer from Berkshire, to a friend, Richard Jarvis, who was at a Vodafone Christmas party. (Since the very early mobile phones of the day did not have the capability to type out individual letters of the alphabet, Papworth had to use his computer keyboard to type out the festive greeting.)

TEXTILES

INGENIOUS Britons transformed the textile industry, creating machines which sped up cloth manufacturing in the 18th century. But behind those inventions is a tale of sabotage and double crossing.

As the industrial revolution took hold, cloth production became ever more important for trade and to clothe the growing population. For generations, cotton or wool had been spun into yarn and then woven into cloth by hand. It was a slow, laborious process until **John Kay**, born in Bury, Lancashire, in 1704, invented the 'Flying

Shuttle' in 1733. It mechanised the process of weaving, allowing thread to be woven into cloth on a loom far more quickly than on a handloom.

The Flying Shuttle was groundbreaking, but financial success eluded Kay. Many manufacturers simply used his machines and refused to pay him the high annual rent he wanted to charge, and hand weavers were none too impressed, fearing that his invention would take their jobs. His house was ransacked in 1753 and shortly afterwards he fled to France where he died a pauper in 1780. (Those who destroyed these new-fangled machines became known as 'Luddites', after a possibly fictional figure called Ned Ludd. The word Luddite has since entered the language as a way to describe those unable to accept new technology.)

Of course, if you increase the speed at which you can weave, you soon need more yarn and spinners were struggling to keep up with the demand. This problem was solved by **James Hargreaves**, a tall, stout man who was born in 1720 in Oswaldtwistle, a small, cotton-producing town in Lancashire. Like most families, the Hargreaves had a spinning wheel in their home, and James realised that it might be possible to spin several threads at once using the same wheel. By 1764, he had built a frame that could spin eight threads. He called it the Spinning Jenny. (It's often said that he named it after his daughter Jenny, except that he didn't actually have a daughter of that name. It's probably a contraction of the word 'engine'.)

Once he started selling the machines, however, there was again a backlash from local workers. They set fire to a workshop containing 20 new machines and, like Kay, he had to flee. He set up in Nottingham but, while he was undoubtedly a talented, self-taught engineer, he wasn't much of a businessman. He failed to patent his product until 1770, by which time it had been widely copied. Improvements were made to his design, increasing the number of threads to 80, and by the time he died in 1778 over 20,000 of them were in use – though they were soon to be obsolete.

Richard Arkwright was born in Preston, Lancashire, in 1732 to a poor family who could not afford to send him to school. His cousin taught him to read and write and he started work as a barber. By 1762 he had set up in business as a wig-maker but they were going out of fashion and he needed a new profession. He happened to meet a Warrington clockmaker called **John Kay** (not the one who invented the Flying Shuttle). Kay's business partner, Thomas Highs, had designed a Cotton Frame which could make a far stronger and longer thread than the Spinning Jenny. The pair had been hoping to develop it, but were out of funds. Arkwright sneakily befriended Kay and persuaded him to reveal the details of Highs' invention. Cutting Highs out of the picture, and using Arkwright's money, the two men built the machine in secret – enquiries were met with claims that they were building a device to solve the longitude problem. Eventually, they revealed the Frame, much to the fury of Highs. Although Arkwright later lost a patents case brought by Highs, he went on to make a fortune using first water mills and then steam power to drive his ever-more impressive machines. A quarrelsome asthmatic, he worked from dawn until dusk every day, and despite his double-dealing was renowned as a good employer. He built cottages for his workers, allowed them a week's holiday a year and refused to employ children under the age of six. These ideas were considered dangerously liberal at the time. When he died in 1792, he had a fortune estimated at £500,000 (£50 million today).

There were many other developments, such as **Samuel Crompton**'s 'Spinning Mule', a mechanised spinning wheel which combined Arkwright and Hargreaves' inventions and produced a finer, softer yarn, and Nottinghamshire-born **Edmund Cartwright**'s 'Power Loom', a steam-powered mechanical engine for making cloth. (His premises were burned down shortly after he received an order for 400 looms – the Luddites again.)

The textile industry was filled with secrecy and intrigue and it ought to be said that some historians suggest Arkwright helped Highs build his original Cotton Frame, or that Kay did more than simply pass on another's secrets, and that Highs built the Spinning Jenny, not Hargreaves.

THE THERMOS FLASK

WHEN **Sir James Dewar** invented **the world's first vacuum flask** in 1892, he wasn't looking for a means of keeping his tea warm on picnics. The Scottish physicist from Kincardine-on-Forth created it as a way of keeping liquids and gases at extremely low temperatures for his experiments. It was expensive and time-consuming to repeatedly liquefy gases, and his 'Dewar' flask – two containers, one inside the other, with a vacuum in between and silvering inside – was designed to minimise heat gain (or loss). It did the job very well, but Dewar did not benefit from his creation. He failed to patent it and, although credited with its invention, there was nothing he could do to stop Thermos from selling flasks for commercial use from 1904 onwards. Incidentally, Dewar (1842-1923) also co-invented a smokeless replacement for gunpowder called cordite, used around the world for many years in munitions, with the London chemist Sir Frederick Abel.

THE THERMOSTAT

ANDREW Ure, a cheesemonger's son from Glasgow, **gave the world the thermostat**.

A doctor and chemist born in 1778, Ure was interested in steam-driven machines and the rapidly-expanding factory system. Textile mills needed a regulated temperature, but this was difficult to achieve, so he invented the bimetal thermostat, patenting it in 1830. It used two lengths of different metals, with different heat properties, fixed together to form a strip. As they heated up, one metal would expand more than the other, forcing the strip to bend. This action would cut off the energy supply, automatically controlling the temperature. As it cooled down it would bend back, switching the heat back on.

The invention led to greater quality goods being produced in British mills.

Ure, who successfully divorced his wife for adultery with another anatomist, is also famous for his gruesome experiments in 'galvanisation'.

Matthew Clydesdale, a weaver who had murdered a 70-year-old man in a drunken rage, was hanged in Glasgow on November 4, 1818. The corpse was delivered to the anatomy theatre at the city's university shortly afterwards, and a large crowd – many of whom had just watched the execution – gathered to see Ure attempt to 'galvanise' the dead man. Using a huge battery, he gave Clydesdale's body a series of electric shocks, leading it to twitch in a ghastly approximation of life. In his notes, Ure wrote that the corpse displayed 'most horrible grimaces… rage, horror, despair, anguish and ghastly smiles… several spectators were forced to leave from terror or sickness, and one gentleman fainted'.

He failed to bring the murderer back to life, though a section on the BBC's website dealing with 'Historical Oddities of the Enlightenment and Industrial Revolution' repeats the myth that he did manage this impossibility. '(T)he corpse was suddenly brought back to life by an electric shock,' it says 'Clydesdale stood up and looked at the professor. Not in the least disturbed, Professor Jeffrey (another anatomist) took out a lancet and plunged it into the bewildered man's jugular vein, who fell on the floor "like a slaughtered ox on the blow of a butcher".' It was galvanisation experiments like this one which inspired Mary Shelley to write *Frankenstein*.

Modern-day doctors can now restart a stopped heart in some circumstances, of course. But **the first person brought back to life** by electrical current was Hannah Sheets, a girl from London. A *Times* report from September 1838 tells how she fell headfirst in a water butt outside her parent's home. A doctor tried to revive her by compressing her chest but was unsuccessful until he 'passed shocks gently through the head and chest, along the course of the spine, gradually increasing their power'. After a short time, 'faint traces of respiration were observed, and in three-quarters of an hour he had the pleasure to behold his patient in a fair way of recovery. The child is now in the enjoyment of perfect health'.

THE THESAURUS

AS a boy, Peter Roget liked to make lists – perhaps as a means of escapism, after the early death of his father, a Swiss-born clergyman, and then the traumatic experience of seeing his uncle commit suicide, by slashing his own throat, when he was only eight years old.

Born in London in 1779, he became a doctor (and was one of the founders of the Royal Society of Medicine, and a secretary of the Royal Society itself) but on his retirement in 1840 he devoted himself to the project which was to define his life. In 1852, aged 73, Roget completed **the world's first 'thesaurus'**, a dictionary of synonyms which he succinctly called *Thesaurus of English words and phrases, classified and arranged so as to facilitate the expression of ideas, and assist in literary composition*. A thousand copies, containing 15,000 words and selling at 14 shilling each, were originally published. It was a huge hit, reprinted 20 times during his lifetime (he died aged 90), and although revised many times it has never been out of print. More than 32 million copies have sold worldwide.

THE TIN CAN

DURING Britain's great age of exploration, hunger and scurvy plagued sailors and explorers and a method of preserving food for long voyages was urgently required. Very little is known about **Peter Durand**, the merchant who patented **the world's first tinned food**. He received his patent from King George III for his proposal in 1810 and promptly sold it for £1,000 (£650,000 today) to Bryan Donkin, a Northumberland engineer, and his business partner John Hall. They set up **the world's first canning factory** in Bermondsey, London, and by 1813 they were supplying tinned meat to the British Army and Royal Navy. The Navy used 24,000 large tins of meat and vegetables on its ships each year by 1818. The cans were made of iron, coated with tin to stop them rusting, filled and then soldered closed with lead. They had to be hammered open because the tin opener had not yet been invented.

THE TOASTER

A YORKSHIREMAN with a handlebar moustache and a mania for electricity, **Rookes Evelyn Bell Crompton** is famous as the man who introduced electricity to India and installed electric lights in Buckingham Palace. But he is most fondly remembered around the breakfast tables of Britain for inventing **the world's first electric toaster**.

Crompton, born near Thirsk in 1845, was obsessed with machines and engineering. He was just 15 when he built a steam-driven road locomotive named 'Bluebell' from scratch. In 1878, he set up his own business, Crompton and Co, which made generating and lighting systems, in Chelmsford. He manufactured the first toaster in 1893. It only toasted one side of bread at a time and the toast didn't pop up when finished, but it certainly beat standing in front of a roaring fire with a toasting fork.

TRAFFIC LIGHTS

THE world's first traffic lights were installed at the intersection of Great George Street and Bridge Street near the Houses of Parliament in December 1868. In what seems like an early example of our national mania for health and safety, they were put up 30-odd years *before* there were even any cars on the road. But in fact mid-Victorian London's roads were actually far more dangerous than today's – at least, according to Westminster Council. The council's website claims that 1,102 people were killed and 1,334 injured on the capital's roads in 1866.

Invented by Nottingham engineer **John Peake Knight** (1828-1886), the signals were operated by policemen. By day, the officers would use a lever to move four-foot-long arms to direct the horse-drawn traffic; by night, they would use gas-powered lanterns – coloured red for stop and green for go.

Ten thousand leaflets were printed to tell Londoners how they would work. Unfortunately, a gas leak caused the traffic lights to explode just three weeks after their installation, burning the face of the police officer on duty, and they were dismantled. Knight also introduced the emergency cord in train carriages. Electrically-powered traffic lights were introduced in the 1920s.

THE TRANSATLANTIC CABLE

IN AN era where it's possible to telephone a friend in Australia while walking in the Cotswolds, it's easy to forget the age before instant communication.

The commercial development of the telegraph in the 1840s meant that, for the first time, you could send messages to distant friends or colleagues. The question was: How far could you send them? The first challenge was to use it to communicate between countries – and next came continents.

In August 1850, **John Watkins Brett** (1805-1863) laid the first line across the English Channel to connect Britain with France. This was **the world's first undersea cable** – a simple copper wire coated with gutta-percha rubber. Bristol-born Brett was a keen art collector who funded the project by auctioning off many of his paintings, raising nearly £7,000 towards the costs. By 1852, Great Britain and Ireland were linked and the following year we were connected to Holland by cable under the North Sea.

The big prize, though, was a link under the Atlantic Ocean to connect Britain and the USA. In 1857, the first attempt to lay this cable started. It was a joint UK-USA project, led by Brett and New Yorker Cyrus Field, and it was a massive undertaking.

The first copper and iron wire cable, made by two English firms from Greenwich and Liverpool, was 2,500 nautical miles long. Insulated with 300 tons of gutta-percha and tarred hemp, it weighed around one ton per mile, and it took three weeks just to load it aboard the two ships which were to lay it between the west of Ireland and Newfoundland. The plan was for the *USS Niagara* to lay her half of the cable first. The *HMS Agamemnon* would then take over in mid-Atlantic, splice on the second half, and complete the task. Amid heaving seas, however, the cable snapped 255 miles out to sea, and the project was abandoned for the year.

In 1858, the two ships tried again. This time they met at the halfway point, spliced their cables and steamed off in opposite directions. After six days at sea, both ships had reached shore and the link was in place. On August 16, Queen Victoria sent the first public message through the cable to President James Buchanan. It took 16½ hours to transmit – a dramatic improvement on the several days a surface-borne message took to arrive, but still quite slow. Despite this, there was widespread delight and excitement on both sides of the Atlantic, and 400 messages were sent. One of these was a signal from the War Office in London cancelling an earlier order for soldiers to return from Canada. That saved British taxpayers the £50,000 cost of transporting the regiment, making the case for the cable very clear.

Unfortunately, within three weeks or so, it was kaput – the cable was fried when the power was increased in an effort to speed up transmission.

In 1865, with modifications suggested by **Lord Kelvin**, who had been involved in planning the operation, a further cable was laid, this time by Isambard Kingdom Brunel's steamer the *Great Eastern* – only for *that* one to snap after 1,062 miles had been paid out.

It wasn't until 1866 that a further cable was *finally* successfully laid and communications re-established.

Later that year, the *Great Eastern* recovered the lost 1865 cable from the seabed two and a half miles below, amid much celebration and cheering. Then they promptly dropped it again, and spent a further two weeks trying to re-find it. When they did, it was connected to a fresh line, meaning there were now *two* links to the USA.

Whereas the 1858 cable had transmitted only one character every two minutes, the 1866 version could send eight *words* a minute – over 50 times faster. Sadly, John Brett had not lived long enough to witness this eventual success. He had gone mad since starting the project, and died in a Staffordshire lunatic asylum in 1863.

Thanks to work by the self-taught English electrical engineer **Oliver Heaviside** (1850-1925), by the 20th century message transmission speeds over transatlantic cables had reached 120 words per minute. This made Britain *the* world centre for telecommunications, with cables radiating out from Porthcurno Cable Station near Land's End to form a girdle right around the world.

The first transatlantic *telephone* cable was laid during 1955 and 1956 and began operating on September 25, 1956. Called TAT-1, it was made possible by (among other things) Heaviside's much earlier invention of the coaxial cable and the creation in the 1930s of practical polyethylene by British scientists at ICI. The 'coax' offered suitable transmission properties and the polyethylene was a better insulator than anything previously used. TAT-1 carried 588 calls between London and the USA and 119 between London and Canada in its first 24 hours.

THE TUNING FORK

LONDON trumpeter and lutist **John Shore** invented the tuning fork in 1711. Still used today, a tuning fork works because when struck its two prongs vibrate uniformly, producing a note of perfect pitch. The note created depends on the length of the prongs. It allows musicians to 'tune' their instruments, and an orchestra to play in harmony.

TYPHOID

FOR centuries, typhoid was one of the great killers, and in some third world countries it still is – the World Health Organisation says 600,000 people die from it each year.

Two Britons **discovered what causes typhoid and how to prevent it**.

A feverish disease which spreads by like wildfire in unsanitary conditions, it's caused by the bacteria *Salmonella typhi* which are excreted by sufferers and passed on when inadvertently ingested. For many years, though, it was thought to spread by the usual 'bad air'.

In 1838, a country doctor from Bristol called **William Budd** (1811-1880) was treating an outbreak in a village in the Taw Valley when he realised that it spread via human contact. If you introduced isolation, you cut cases dramatically. But his insights were largely ignored and it wasn't until 1873 that people began to take him seriously. (Even then, not everyone agreed with him; until 1900, the water supplies in Paris were often topped up with raw Seine water, leading to repeated typhoid outbreaks.)

Unfortunately, typhoid has an incubation period of up to 20 days, and some people carry it without ever becoming infected. Thus, it could never be properly eradicated without a vaccine (see *Vaccination*). **Almroth Wright** (1861-1947) was born the son of a 'violently anti-Papist' Ulsterman at Middleton Tyas, Yorkshire, in 1861. He is described by his biographer Michael Dunnill as a passionate anti-suffragettist who believed that there was no place for women in medicine, and as possessing 'an overwhelming arrogance' and an astonishing memory (he was said to know a quarter of a million lines of poetry). He was also a brilliant early immunologist. At the end of the 19th century, British soldiers were far more likely to die of typhoid than at the hands of any enemy. Wright developed **the world's first typhoid vaccine** in 1896 at the Army Medical School in Netley, Hampshire, injecting 'dead typhoid' (heat-treated bacteria) into himself and finding that it conferred immunity. British troops were vaccinated at the outbreak of WWI and it was later calculated that up to half a million lives were saved by its use. The French army, which did not vaccinate against typhoid, suffered many thousands of cases, while the British figures were in the hundreds.

Wright also saved military lives with a change in the way soldiers with battlefield wounds were treated. Following the work of Joseph Lister, army medics packed wounds with antiseptic. Wright believed that, in order to reduce the risk of infection and gangrene, it would be better to thoroughly clean them, stitch them up, insert a drainage tube and provide a sterile saline drip. His ideas were slow to be adopted but later became standard practice.

Each year, between 150 and 200 cases of typhoid are still diagnosed in England and Wales. Unfortunately, as Wright predicted they would, the bacteria are developing immunity to antibiotics.

THE TYPEWRITER

THE Americans like to claim that they invented the typewriter, but a London engineer came up with the idea more than a century earlier. In 1714, **Henry Mill** (1683-1770), who worked for the New River Company which transported fresh water from Hertfordshire to the capital, was granted a patent for **the world's first typewriter**. He described it, rather snappily, as 'An Artificial Machine or Method for the Impressing or Transcribing of Letters, Singly or Progressively one after another, as in Writing, whereby all Writing whatsoever may be Engrossed in Paper or Parchment so Neat and Exact as not to be distinguished from Print'. It's not known whether he ever made his machine, however. The first US patent was granted in 1829.

TYRES

THE **world's first pneumatic tyre** was invented by a Scotsman. You might assume that that Scotsman was **John Boyd Dunlop**, but the tyre was actually the brainchild of the obscure 11th son of a wool mill owner from Stonehaven in Kincardineshire. **Robert Thomson** (1822-1873) left school at 14 and taught himself chemistry, electricity and astronomy 'aided by a weaver who was a mathematician', according to the *Oxford Dictionary of National Biography*.

Thomson patented his pneumatic tyre in France and the USA in 1846 and 1847. The patent was for hollow belts of India-rubber inflated to provide 'a cushion of air to the ground, rail or track on which they run' and further enclosed in a strong leather outer casing bolted to the wheel. His 'Aerial Wheels' were fitted to several horse-drawn carriages in London in 1847, improving comfort and reducing noise. One set ran for 1,200 miles without sign of deterioration. Unfortunately, it was impossible to obtain sufficient quantities of rubber to make his invention commercial.

Scottish vet John Boyd Dunlop, who was born in Dreghorn, Ayrshire, in 1840, 're-invented' the pneumatic tyre 40 years later, developing it for his son's tricycle and patenting it in 1888. His patent was withdrawn in favour of Thomson's two years later. Dunlop continued making tyres, though his financial benefits were limited when he sold his shares in the company early on. Dunlop, of course, is synonymous today around the world with tyres.

Thomson, incidentally, also came up with the idea of electrically-ignited controlled explosions, designed a self-filling fountain pen, invented the sprung mattress, produced early designs for caterpillar-tracked vehicles and greatly improved on his era's steam boiler designs.

ULTRASOUND

AN ingenious doctor and war hero from Cornwall invented **the world's first ultrasound scanner**.

When **Professor Ian Donald** first suggested using sonar to examine unborn babies and their mothers, his colleagues thought he was crazy. But he pressed ahead, and in 1958 he transformed maternity services around the world.

Donald, a doctor's son born in 1910, served in the RAF as a medical officer during WWII, and was decorated for rescuing injured airmen from a burning, bomb-laden aircraft. Known for his sharp wit and flaming red hair, he had already invented a respirator for newborns with breathing difficulties while working as an obstetric gynaecologist at Hammersmith Hospital in London after the war. In 1954, he moved to Scotland to become chairman of midwifery at Glasgow University. He wondered whether the sonar technology used to locate submarines could be adapted to detect abnormalities in the womb, with high-frequency sound waves reflecting back from the internal tissues to create a picture (now called a sonogram).

The following year, he borrowed an ultrasonic metal flaw detector – used to identify problems with the hulls of boats and armour on tanks – from the husband of one of his patients and tested it on body parts stored in the university laboratory. Hearing that a strange professor was experimenting with this device, a resourceful young Glaswegian, **Tom Brown**, contacted him. Brown, who designed scientific instruments at the Kelvin Hughes Scientific Company, was convinced that Donald was on to something and persuaded his bosses to let him help the professor build a prototype scanner. With very little money for the project – Brown was told he could spend half a day a week on it, with a total budget of £500 – the first was a bizarre mishmash of begged and borrowed bits and bobs, and used sprockets and chains from another British invention, Meccano.

With the first purpose-built device in hand, Donald and Brown enlisted **Dr John MacVicar** and they began the painstaking work of using ultrasound to distinguish the differences between deadly tumours and less serious cysts and fibroids. The trio's early results were disappointing and Donald was ridiculed – until he used ultrasound to detect a huge but easily removable ovarian cyst in a woman who had been told that she had terminal stomach cancer. In 1958, he submitted a paper to the *Lancet* after using ultrasound on 100 patients. He also used it to measure foetal growth, diagnose complications and identify multiple pregnancies. It is now a routine part of pre-natal care, is used in other medical fields and has advanced further with 3D images now available from some scanners.

THE UMBRELLA

THE collapsible steel ribbed umbrella was invented by **Samuel Fox** in 1852. Born in Bradwell, Derbyshire, in 1815, he is said to have been spurred on by the wet weather in the Peak District. Fox's 'Paragon Umbrellas' were more robust and much easier to fold when wet than the wooden ones then in use. At the time, men carrying umbrellas in public were ridiculed and (even worse) accused of being French.

THE UNDERGROUND

ON Saturday 10th January 1863, people jostled and queued for ages to get on a train. So what's new, you ask? Well, it was the opening day of the Metropolitan Railway, which formed the first part of the London Underground – **the world's first underground railway**, and the first to use trains powered by electricity.

An astonishing 38,000 people rode the trains on the opening day, on a journey of 3.75 miles from Paddington to Farringdon St in the City of London. Some passengers complained of overcrowding, overheating and foul smells. (So, no change there then, either.)

The origins of the 'Tube' lay in the street congestion in the London of the 1840s and 1850s – one witness reported that it took longer to get across the capital than it did to travel down to Brighton. (Once again, pretty much like today). Lawyer **Charles Pearson** had campaigned for the underground for years, and work started in 1860 with the Sheffield-born civil engineer **John Fowler** in charge. Sadly, Pearson did not live to see his dream fulfilled: he died of dropsy in September 1862, three months before the Metropolitan line opened.

The first tunnels were shallow affairs, excavated by the cut-and-cover method of digging a trench, putting a roof on top and covering it over with dirt (despite its name, 55% of the present network is actually above ground). Nevertheless, the work cost £1 million (£1.5 billion as a share of GDP today). Today its trains carry a billion passengers (by passenger journeys) a year at an average speed of 20.5mph around the entire 249 mile network.

URANUS

THE planets Mercury, Venus, Mars, Jupiter and Saturn – named by the Romans after their gods – are visible with the naked eye.

In 1781, the British astronomer **William Herschel** (1738-1822) became **the first modern astronomer to discover a new planet**. Herschel was a keen amateur who built telescopes to make observations from his house in Bath. One night, he spotted an object that, from its size and shape, did not look like a star. Over the following nights, he observed it methodically and was able to work out that it was orbiting the Sun, and that its orbit was too circular for it to be a comet. He originally named the planet 'Georgium Sidus' (George's Star), probably in gratitude for a gift of £4,000 from King George III with which he had built his main telescope. The French preferred to call it 'Herschel' but a German named it 'Uranus' (after the Greek and Roman god of the sky) and that stuck, paving the way for a million poor jokes.

Herschel became a full time astronomer and made numerous discoveries, including two new moons around Saturn and Uranus. He also discovered that 'double stars' weren't an optical illusion caused by poor technology, as was believed, but were actually two distinct stars (now known as binary stars) orbiting each other. The importance of this discovery was that it proved that Newton's Law of Gravity worked outside our Solar system and wasn't just a local rule.

He built over 400 telescopes, one of which had a 48-inch diameter and, at 40ft in length, remained the largest in the world for over 50 years. He was also the first to realise that our galaxy is in the shape of a disc and the first to take measurements of the 'Proper Motion' (the actual movement, relative to the Earth) of certain stars. This allowed him to deduce that our whole solar system is moving through space, and to estimate the rough direction in which it is heading. He also discovered infra red radiation whilst doing an experiment to split light into its component colours using a prism. He was measuring the temperature of each colour in turn, and noticed that it increased towards the red end of the spectrum and was even higher just off the end where he could see nothing. This was the first demonstration that there were forms of light beyond our vision. It was a classic accidental scientific discovery by an enquiring mind intelligent enough to spot an anomaly.

(Herschel was actually born 'Friedrich Wilhelm Herschel' in Germany, but moved to Britain as a young man. He died in Slough aged 84 – coincidentally, Uranus takes 84.3 years to orbit the Sun – that is, one Uranian year is 84.3 Earth years.)

VACCINATIONS

IN 1796, an enormous breakthrough in the treatment of diseases was made by a daring and imaginative Gloucestershire doctor called **Edward Jenner**, when he carried out **the world's first vaccination**. Along the way, Jenner also helped to eradicate one of the world's biggest killers, smallpox. This highly-feared and terribly contagious disease killed 400,000 Europeans a year in the 18th century, at a time when the population was far smaller than it is today. Caused by the *Variola major* and *V. minor* viruses, and spread by contact with an infected person, it left those who contracted it covered in boils and, in 30 percent of cases, dead from organ failure, bleeding or sepsis. A third of those who lived were blinded, and all were scarred. (Queen Mary II and Edward VI were among its victims, while famous survivors included Mozart, Beethoven, Abraham Lincoln and Joseph Stalin.)

Jenner (1749-1823), a clergyman's son, was interested in folklore that suggested milkmaids who contracted cowpox – a mild viral infection of cows which can cause weeping spots and a brief feeling of unwellness in humans – rarely caught smallpox. In May 1796, a dairymaid called Sarah Nelmes consulted him about a rash on her hand. He diagnosed cowpox and she confirmed that Blossom, one of her herd, had recently had the disease. Jenner realised that this was his chance to test the theory that cowpox protected against smallpox.

On May 14, he took James Phipps, the eight-year-old son of his gardener, made a few scratches on his arm and rubbed in pus from one of the boils on Sarah's hand. A few days later James became ill with cowpox – proving that it could pass from person to person as well as cow to person.

Then, on July 1 – in a trial which would probably be frowned on today – Jenner tried to infect James with smallpox.

Imagine the trepidation he, the boy and his family must have felt at this - and their great relief when James appeared to be immune. Subsequent attempts to infect him with smallpox were also, fortunately, unsuccessful and he lived to the age of 65.

Jenner's was not an overnight success. He was widely ridiculed, particularly by members of the clergy who said it was ungodly and repulsive to inoculate someone with material from a diseased animal. To test the theory further, he gave the same treatment to 23 other children, including his own 11-month-old son. There were no fatalities and in 1798 his results were published. Jenner had developed **the world's first successful vaccine**. He also coined the word vaccine from the Latin *vacca*, for cow.

In 1967, the World Health Organisation (WHO) estimated there were still 15 million cases of smallpox a year, mainly in South America, Africa and the Indian subcontinent, and launched a worldwide vaccination campaign. The last known case of naturally-acquired smallpox occurred in Somalia in October 1977, though the last victim was Briton Janet Parker, a 40-year-old medical photographer working at the University of Birmingham. She contracted the disease in 1978 when working in a darkroom above a lab where research with live smallpox was being carried out.

In 1980, the WHO formally declared smallpox eradicated, fulfilling a prediction that Jenner had made in 1801. To date, it is the only disease to have been eliminated through an immunisation programme. It has been estimated that Jenner's breakthrough has led to the saving of more human lives than the work of any other person.

THE VACUUM CLEANER

WHEN **Hubert Booth** watched a demonstration of an 'amazing new cleaning machine' for railway carriages, he was unimpressed. Choked by the flying dust that the American inventor was blowing out of a carpet and trying to catch in a box, he asked why he did not use suction instead. Angry with the Gloucester engineer's impudence, the inventor said that it was impossible and stalked out of St Pancras Station in a huff.

But Booth could not rid himself of the idea and, in 1901, he developed **the world's first vacuum cleaner**. His contraption, called the Puffing Billy, was so big that he used a horse-drawn cart to transport it from job to job and it had to remain outside the building while in use, with an 82ft hose threaded through an open window. Powered by a petrol generator, it was bright red and incredibly noisy. Booth, who was born in 1871, said he was 'frequently sued for damages for allegedly frightening cab horses in the street'.

To promote it, he offered to clean the dining room of a local restaurant for free. It gained huge publicity, and one of his first jobs was cleaning the carpet in Westminster Abbey before the coronation of Edward VII in 1902. Portable electric versions were on the market by 1906, and soon his vacuum cleaners were installed in Buckingham Palace, the House of Commons and Windsor Castle. As the machines were so expensive to buy, roughly £350 each (over £25,000 today using an RPI measure), Booth employed operators to go to people's homes for an annual subscription of £13 (£1,000). It made him a wealthy man.

VD CLINIC

IN 1747 the Lock Hospital in London was founded by surgeon **William Bromfield** (1713-1792). It was **the world's first VD clinic**.

Syphilis, particularly, is a very serious condition which can send sufferers mad – there was an asylum attached – or even kill them. Since there were no effective treatments for sexually transmitted diseases until the arrival of antibiotics, Bromfield concentrated on moral reform, in between trying novel but useless ointments and potions on his patients.

VERTICAL TAKE OFF

AERO-ENGINEERS had dreamed of Vertical/Short Take Off and Landing Aircraft for decades but they all failed until **Alan Arnold Griffith** (1893-1963) joined Rolls-Royce in Derby as a research engineer. Obsessed with engines and aerodynamics, London-born Griffith was a tall, slim, serious man who shunned the limelight. His ingenious designs led to the production of a testing platform called the Thrust Measuring Rig, nicknamed 'The Flying Bedstead' because of its unusual appearance – it was a metal framework which supported two Rolls-Royce engines with smaller jets for lateral control. **The world's first vertical flight** took place at Hucknall, Nottinghamshire on August 3, 1954, with Rolls-Royce's chief test pilot Captain Ron Shepherd, at the controls. It was a noisy and hair-raising experience, with the machine hovering 10ft above the ground for eight and a half minutes. This led indirectly to the development of Britain's Harrier 'jump jet'. This iconic aeroplane – the only successful VSTOL aircraft yet built – is still in service with the Royal Air Force and the US Marine Corps 40 years after it was first introduced. It played a large part in the defeat of Argentina during the 1982 Falklands War.

VIAGRA

FOR centuries, men have sought outside assistance to stiffen their bedroom resolve. In ancient Rome and Greece, eating the genitals of roosters and goats were recommended as a cure for impotence. Albertus Magnus, a 13th century friar, claimed roasted wolf penis was an aphrodisiac. In 1922, a doctor at California state prison in San Quentin claimed to have implanted the testicles of executed convicts into men with erectile dysfunction. Other treatments included vacuum pumps, penile implants and medicine injected directly into the root of the problem.

A simpler and far more palatable solution was provided by British scientists. In 1991, **Andrew Bell**, **David Brown** and **Nicholas Terrett** were working on a drug for heart problems at a Pfizer research centre in Kent. They patented 'Sildanafil' and clinical trials were started. Pretty quickly they found that the drug was not particularly good for treating angina – but it did have an interesting side effect. It increased blood flow to the penis in male patients, causing erection. In March 1998 it was approved as **the world's first impotence cure** and the little blue pill – renamed 'Viagra' – became the fastest-selling new drug in history.

THE VICTORIA CROSS

WE didn't invent battlefield bravery, or medals, but we *have* given the world hundreds of examples of incredible courage. They are marked by our highest award for heroism, the Victoria Cross. It was introduced on the orders of Queen Victoria as a way to commemorate acts of ludicrous gallantry during the Crimean War. It was the first 'classless' medal, in that it was awarded regardless of rank and is given for acts of supreme bravery in the presence of the enemy.

Here are three astonishing tales, one from each service.

Charles Lucas (1834-1914) was awarded the first ever VC in June 1857 for an action that had taken place three years earlier, at the start of the Crimean War with Russia. A 20-year-old Royal Navy

Midshipman from County Armagh, his paddle-wheel steam ship, *HMS Hecla*, was one of three vessels which attacked the fortress of Bomarsund, on an island off the coast of Finland. In typical British fashion, we were vastly outgunned and the mission appeared almost doomed to failure from the start. The *Hecla* had just eight guns and the *Odin* and *Valourous* 16 each, while the well-defended, granite-built fortress had more than 100. The Victoria Cross Society describes the attack as 'foolhardy'. Still, our sailors began an enthusiastic bombardment. Before long, the fortress began replying in kind, and a live shell hit the *Hecla* but failed to explode. As it lay fizzing on the deck, its fuse clearly burning down, everyone dived for cover – except Lucas. He sprinted across the deck to the shell, picked it up (it would have been red hot) and threw it overboard. It exploded before it even hit the water.

His action – under constant fire from the fortress – saved dozens of lives. If the shell had blown up on the deck, the shrapnel would have killed many of the sailors nearby and possibly sunk the ship. His Captain, William Hall, promoted him to Lieutenant on the spot (Hall was later given a stern rebuke from the Admiralty for recklessly endangering his ship and wasting all his ammunition, though the public back home loved the audacity of the raid).

Lucas remained in the Navy for another 15 years, retiring as a Captain to live in the Highlands. In a bizarre twist, he was later called to the death bed of his old Captain, where Hall begged him to marry his daughter. Lucas agreed to do so, though he later regretted it as she was what we might today call a 'nightmare'. Still, they moved to Tunbridge Wells and had three children. Lucas was promoted to Vice Admiral on the retired list, and died in 1914, at the age of 80.

Now, if that's the sort of thing that wins you a VC, what sort of crazy fool wins *two* of them? In fact, three men hold this extreme honour, including **Noel Chavasse**. Both were awarded during the First World War. Chavasse, a twin born in 1884, was very small and weak at birth, and almost died of typhoid in his first year. He grew up in Oxford and, later, Liverpool, after his father was made Bishop of that city. In 1912 he qualified as a doctor and joined the Royal Army Medical Corps. On the outbreak of the war, he was sent to France with his Battalion. His first medal was the Military Cross, awarded after the savage Battle of Hooge in June 1915, at which only 140 of the 900 men in his unit were left standing. This in itself is a very serious gallantry award, though the citation detailing his actions is unfortunately lost. He was also further Mentioned in Dispatches for his bravery in November that year.

His first VC was won at Guillemont on the Somme on August 9, 1916. This was typical of many small battles of the war – it involved the pointless sacrifice of hundreds of lives in order to take a small and unimportant piece of land which would later be surrendered to the enemy when they launched their own suicidal advance. The British soldiers went forward under sweeping machine gun fire, with shells exploding around them, stepping over the bodies of friends and comrades who had fallen in previous attacks. Hundreds were killed and injured, and Chavasse – now a Captain – spent the day and night under heavy fire, often in plain view of the Germans, recovering the wounded. He saved the lives of some 20 badly-wounded men, and was himself injured by shell splinters as he carried stretchers. His citation says: 'His courage and self-sacrifice were beyond praise.'

His second VC came for his actions between July 31 and August 2, 1917, at Wieltje, Belgium. Pushing forward with an attack, he was hit in the head by shrapnel, possibly suffering a fractured skull. Against advice, he stayed in the field. Ignoring enemy bullets, practically without food, worn with fatigue and faint from his injury, he crawled in the pouring rain through the mud and gore in search of wounded men until well into the night. He helped to carry in many who would have died but for his bravery. The following day, he suffered further head injuries as shells exploded all around his medical dugout. Despite the intense pain, still he refused to leave. Then, in the early hours of August 2, a shell hit his aid post, killing or badly injuring every man inside. Chavasse's abdomen was ripped open by steel shards from the explosion. He managed to crawl out of the dugout and along the sodden, mud track to another aid post. From there he was taken to a Casualty Clearing Station at Brandhoek, and operated on immediately. Although he regained consciousness, he succumbed to his wounds on August 4. Incredible stuff.

Flight Sergeant **Norman Jackson** (1919-1994) won *his* VC for a piece of ridiculous bravery high in the night sky over Germany on April 26-27, 1944. Ealing-born Jackson flew a 'tour' of 30 missions as a flight engineer in an Avro Lancaster bomber with 106 Squadron, which earned him the right to a break from operations. But he volunteered for one more mission, a bombing raid on ball bearing factories at Schweinfurt in central Germany. Jackson's Lancaster successfully bombed the target, but as the pilot turned for home they were attacked by a German fighter and a fuel tank in the starboard wing caught fire amid a hail of bullets. Jackson – already wounded from anti-aircraft shrapnel – quickly formed a plan. He would open his parachute in the cockpit, and the navigator and bomb aimer would pay out the chute's guide lines, as a sort of safety rope, as he climbed out onto the wing with a fire extinguisher.

As his citation makes clear, this was virtually a suicide mission: 'This act of outstanding gallantry… was an almost incredible feat. Had he succeeded in subduing the flames, there was little or no prospect of his regaining the cockpit.'

Ignoring the risks to his own life, Jackson jettisoned the escape hatch and clambered outside. His aircraft was 20,000ft up in the freezing night sky, roaring along at 200mph, with the enemy still somewhere nearby. He slid down onto the wing and began inching himself along it to the blaze. As he reached the fuel tank, he started to extinguish the flames, sustaining severe burns to his hands and face. But just as he started to control the fire, the German fighter re-attacked.

The British pilot was forced to bank hard, and Jackson was hurled through the fire and off the wing, his parachute, partially inflated and burning, trailing behind him. His comrades watched in horror as he fell. With his chute badly damaged, he plummeted to earth in an uncontrolled descent. Amazingly, he fell into thick bushes and survived to be picked up and imprisoned as a POW by the Germans. He had sustained two bullet wounds to his leg and a broken ankle to add to his flak injuries and burns. (The Lancaster was abandoned in mid-air shortly after his exploits, with the Captain and tailgunner dying but the other four crew, like Jackson, being caught by the enemy.)

After 10 months in captivity – during which he did not mention his actions once – Jackson managed to escape from his prison camp to freedom. It was only after the war ended, when his comrades were released, that they told of his amazing bravery. He was instantly recommended for a Victoria Cross. On leaving the RAF, he worked as a salesman for a whisky distiller before dying in 1994. He rarely talked about his exploits and seldom wore his medals. 'It was almost as though they were an embarrassment,' said his son, David. 'He didn't think he'd done anything out of the ordinary.'

VITAMINS

NOBEL Prize winner **Sir Frederick Gowland Hopkins** started work as an insurance clerk, but the bookseller's son from Eastbourne went on to make an astonishing scientific breakthrough in the field of nutrition.

In 1906, Hopkins (1861–1947) proved that a diet rich in calories was not enough for animals to thrive and that 'accessory food factors' – what we now call vitamins – were essential for health and growth. He qualified as a doctor aged 32 and taught at Guy's Hospital for several years before moving to Cambridge University in 1898 to lecture and carry out research into the nutritional value of food.

Hopkins fed mice basic protein, fat, starch and salts and found that they stopped growing. As soon as he added milk to their diet they developed normally. During further experiments he realised that these 'accessory food factors' occurred in fresh foods and in very small quantities. Academics were sceptical at first. They believed that diet-linked illnesses were caused by toxins within foods and that, as long as an animal received enough calories, it would grow. Later scientists went on to identify individual vitamins. Hopkins became Cambridge's first Professor of Biochemistry in 1914, was knighted in 1925 and won a Nobel Prize in 1929.

WATERBEDS

ANXIOUS to help bed-bound patients in hospitals, a Scottish doctor came up with the first waterbed to prevent bed sores. **Neil Arnott** (1788–1874), from Arbroath, realised that a water-filled mattress would distribute weight evenly and could stop painful pressure sores. He made a prototype in 1832 that was used in St Bartholomew's Hospital, London. It did not catch on because it leaked and the water temperature could not be controlled so it was too cold to sleep on.

WEATHER FORECASTS

LEWIS Fry Richardson (see *Sonar*) revolutionised weather forecasting while working as a WWI ambulance driver in France.

Richardson studied science at Cambridge and was employed by the Meteorological Office in Scotland when the war broke out in 1914. As a Quaker, he became a conscientious objector, but he did ferry wounded soldiers on the battlefields. During this time he wrote a book called *Weather Prediction by Numerical Process*. It was published in 1922.

At the time, meteorologists looked for similar weather patterns from old records and made their predictions based on what had happened next. Richardson was the first scientist to use mathematics to make forecasts. He theorised that observations from weather stations would provide raw data which could be used to calculate future conditions by specific equations. Unfortunately, the observatories of the day did not record enough data, and the computers – still, at that time, people who 'computed' – could not process it quickly enough anyway.

According to the *Encyclopaedia Britannica*, 'The main drawback to his mathematical technique for systematically forecasting the weather was the time necessary to produce such a forecast. It generally took him three months to predict the weather for the next 24 hours.' Richardson calculated that he would need 60,000 people to do the sums and solve the equations necessary to predict tomorrow's weather before it actually arrived. It was the 1950s before the early mechanical computers were in place to make use of his methods.

WELLIES

WELLINGTON boots were created nearly 200 years ago, when the Duke of that name, the stylish war hero Arthur Wellesley, instructed his cobbler to come up with something different. The shoemaker, who worked in St James Street, London, designed a snug-fitting, knee-length boot that could be worn under trousers. Wellington was delighted, and the boots became so popular that they were named after him. They were originally made out of calf-skin – the process for using rubber to make boots had not yet been invented – and they replaced Hessian boots which were wider around the calf and had a decorative tassel at the top.

In the 1840s, the American Charles Goodyear found that, by adding sulphur to rubber and then heating it, it would keep its elasticity. This process was called 'vulcanisation' and it opened the door for more products to be made out of rubber. These included the first rubber wellies, and most sources agree these were first produced in Scotland by an American tycoon, Henry Lee Norris, who set up the North British Rubber Company at the Castle Silk Mills in Edinburgh and began production in 1856.

WHISTLES

IN 1883, the Metropolitan Police were looking to find a replacement for the wooden rattles they used to summon help. These were cumbersome, and their noise didn't penetrate far in the dank and foggy back streets of the capital. **Joseph Hudson** was originally from Derbyshire, but had moved to Birmingham and trained as a toolmaker. He designed the 'Thunderer', the first whistle to use a 'pea'. It gave a very distinctive sound which carried much further than rivals and, when tests on Clapham Common showed that it could be heard from up to a mile away, the police invested. Hudson called his trading company 'Acme' from the Greek for 'pinnacle' and went on to design whistles for football referees (who previously had waved handkerchiefs), locomotives and dog owners. ACME supplied the whistles and megaphone for the *Titanic* and still sells five million products a year worldwide.

WINTER SPORTS

STRANGELY, given that Britain is a relatively flat country and we don't get much snow, we have been heavily involved in winter sports. Ski-ing originated in northern Europe around 7,000 years ago, but it was a Briton, **Arnold Lunn** (part of the Lunn Poly travel agency dynasty, born in 1888), who invented slalom ski-ing (in 1922 in Mürren, Switzerland) and introduced downhill and slalom racing into the Olympics in 1936.

The world-famous Cresta Run – a terrifying, head-first, 80mph plunge on a skeleton sledge down a 3,978ft course in the Swiss hamlet of Cresta – was created in 1884 by a British Army officer called **Major W H Bulpetts**, and bobsleighing was invented by English holidaymakers in St Moritz, Switzerland in 1890.

The slightly less petrifying game of curling – effectively bowls played on ice with flat stones – originated in Scotland (the first club was formed at Kilsyth, Stirlingshire, in 1510); curling hysteria briefly swept the UK in 2002, when Rhona Martin and her colleagues in the GB team won gold at the Salt Lake City Winter Olympics. Ice hockey is another British invention – most people believe that the game was first played by British soldiers staying in Nova Scotia, Canada, in the 1850s.

WOMEN'S LIBERATION

ASTONISHING as it now seems, women did not always have the right to vote in general elections (or – as we've seen – graduate from universities, or become doctors, or do many other things men could). Although New Zealand was the first self-governing country where women won the vote, Britain's 'suffragettes' became the leading standard-bearers of a worldwide movement.

Among the most famous of them was **Emily Davison**, a merchant's daughter from Greenwich. She died, aged 40, when she ran out in front of King George V's horse in the 1913 Epsom Derby. There is debate as to whether she intended to commit suicide (she had bought a return train ticket to the event), or was trying to unfurl a protest banner or perhaps just disrupt the race. But her death made front page news around the globe and undoubtedly helped to hasten women's suffrage.

WRISTWATCHES

THE world's first automatic wristwatch was invented in 1923 by **John Harwood**, a watch repairer from the Isle of Man. It wound itself with the movement of the wearer's arm, and went on sale in 1928; 30,000 were made, only for Harwood's business to collapse in the Great Depression of 1931.

WWW

WE take it completely for granted now, but it was only in 1989 – whilst working at CERN (The European Centre for Nuclear Research) – that **Tim Berners Lee** designed the world wide web, a system that enables documents to be shared between computers. If he'd wanted, Berners Lee could perhaps have patented the idea and charged a small royalty on every downloaded page or email. Instead, he made his work freely available.

He was born in 1955 and grew up in London in a computer-obsessed family. His parents, Conway Berners Lee and Mary Lee Woods, had worked on the development of a very early computer, the 'Mark 1', at Manchester University in the mid-20th century, and he was himself temporarily banned from using his university computer after being caught hacking. He is now living in America and working on his next project – the 'Semantic Web', which he hopes will make data on the internet yet more accessible and easier to use, with complex and specific searches for information being catered for.

Also from Monday Books

Wasting Police Time / **PC David Copperfield** (ppbk, £7.99)

The fascinating, hilarious and best-selling inside story of the madness of modern policing. A serving officer - writing deep under cover - reveals everything the government wants hushed up about life on the beat.

'Very revealing' – *The Daily Telegraph*
'Passionate, important, interesting and genuinely revealing' – *The Sunday Times*
'Graphic, entertaining and sobering' – *The Observer*
'A huge hit... will make you laugh out loud'
– *The Daily Mail*
'Hilarious... should be compulsory reading for our political masters' – *The Mail on Sunday*
'More of a fiction than Dickens'
– **Tony McNulty MP, former Police Minister**
(On a BBC *Panorama* programme about PC Copperfield, McNulty was later forced to admit that this statement, made in the House of Commons, was itself untrue)

**From all good bookshops, online from
www.mondaybooks.com or via 01455 221752.**

Perverting The Course Of Justice / **Inspector Gadget**
(ppbk, £7.99)

A senior officer picks up where *Wasting Police Time* left off. A savage, eye-opening journey through our creaking criminal justice system, which explains what it's really like at the very sharp end of British policing.

'**Exposes the reality of life at the sharp end**'
– *The Daily Telegraph*

'**No wonder they call us Plods… A frustrated inspector speaks out on the madness of modern policing**'
– *The Daily Mail*

'**Staggering… exposes the bloated bureaucracy that is crushing Britain**' – *The Daily Express*

'**You must buy this book… it is a fascinating insight**'
– **Kelvin MacKenzie**, *The Sun*

In Foreign Fields / **Dan Collins**
(ppbk, £7.99)

A staggering collection of 25 true-
life stories of astonishing battlefield
bravery from Iraq and Afghanistan...
medal-winning soldiers, Marines
and RAF men, who stared death in
the face, in their own words.

'Enthralling and awe-inspiring untold stories'
– *The Daily Mail*

'Astonishing feats of bravery illustrated in laconic,
first-person prose' – *Independent on Sunday*

'The book everyone's talking about... a gripping account
of life on the frontlines of Iraq and Afghanistan'
– *News of the World*

'An outstanding read' – *Soldier Magazine*

It's Your Time You're Wasting / Frank Chalk
(ppbk, £7.99)

The inside story of real life in a modern British comprehensive school - as told by teacher Frank Chalk. It veers from tragedy to farce... drunk pupils, drugs and porn in the classroom, attacks on staff - and a few brave kids who dare to swim against the tide.

'**Ghastly and addictive**' – *The Times*

'**One of the three best political books of the moment**' – **Nick Cohen**, *The Observer*
(*Wasting Police Time* was one of the other two)

A Paramedic's Diary / **Stuart Gray**
(ppbk, £7.99)

Stuart Gray is a paramedic dealing with the worst life can throw at him. A Paramedic's Diary is his gripping, blow-by-blow account of a year in on the streets – 12 rollercoaster months of enormous highs and tragic lows. One day he'll save a young mother's life as she gives birth, the next he might watch a young girl die on the tarmac in front of him after a hit-and-run. A gripping, entertaining and often amusing read by a talented new writer.

**From all good bookshops, online from
www.mondaybooks.com or via 01455 221752.**

Not With A Bang But A Whimper / **Theodore Dalrymple**
(hbk, £14.99)

Prison doctor, psychiatrist and cultural commentator Theodore Dalrymple - in a series of penetrating and beautifully-written essays - explains why he thinks the liberal intelligentsia are destroying Britain. Dalrymple writes for *The Spectator*, *The Times*, *The Daily Telegraph*, *New Statesman*, *The Times Literary Supplement* and the *British Medical Journal*.

The Curse of the Al Dulaimi Hotel / Colin Freeman

(ppbk, £7.99)

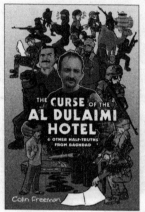

Sunday Telegraph foreign editor Colin Freeman on the tragedy of Iraq. A moving, witty and warm appreciation of the Iraqi people and the quaqmire created in their country by George Bush and Tony Blair.

AS KIDNAPPED BY SOMALI PIRATES!

'An amusing book... Freeman is the quintessential "little man", alone on the spot. Humorous and self-deprecating' – *Daily Mail*

'Freeman's light tone darkens as the violence in Baghdad escalates – he communicates his experiences in a refreshingly unmacho way' – *Sunday Telegraph*

From all good bookshops, online from www.mondaybooks.com or via 01455 221752.